INVITATION

Dear Reader,

The year 2020 has been a special year; so different than expected.
We all have a few of our own 2020 stories.

At 9:37 am CET on November 28, 2020, WHO reported that there have been
globally 61,036,793 confirmed cases of Covid-19 and 1,433,316 deaths.[1]
These are numbers. Each represents a person, an individual story, and
numerous other people, circumstances and systems involved.

We are still right in the middle of the pandemic and do not yet know
where it will lead—neither in terms of the disease's development nor the
lives of countless people. It also remains to be seen what the long-term
consequences of this pandemic will be on businesses and the global
economy. These are times of disruption, uncertainty and change.

Why did I start a book about sustainability with data about a virus?
For two reasons.

The first is very personal. I started to write the initial pages of this book
just before the first lockdown hit Europe in March 2020. With the world
shaken and many people devastatingly affected, I found myself despera-
tely thinking many times "Heavenly Father, where are you? Have mercy
on us!" What I have since learned is that the God I believe in is indeed
present and active during this crisis, but in ways other than I had thought.

During 2020, Covid-19 superimposed itself on sustainability. But, while
the coronavirus will pass, the challenges related to sustainability will
remain. Working in this field for almost 20 years, I'm familiar with a wide
range of topics. Still, while writing the book, I was moved when I saw how
dire some situations really are and how urgently action is needed. These
next ten years will be absolutely decisive for sustainability.

This book and its topics have been my constant companion these past
several months and writing it has inspired me in these turbulent times.
To be able to dedicate my talent and energy in this period of change and
create something that will have implications beyond this season, gave me

personal hope. My wish and my prayer are that the knowledge and encouragement in this book will spread and have an impact on some people in the future in order to multiply positive effects.

The second reason: Sustainability has some astonishing parallels to Covid. US author and environmentalist Stewart Brand stated *"the climate crisis will be more expansive, more dangerous and more deadly than the actual pandemic ... but it is a catastrophe that sneaks up in slow motion and misses the momentum of the total suddenness which caused immediate reaction."* [2]

If we open ourselves up to this perspective, Covid-19 can serve as a wakeup call. Climate change, loss of biodiversity, resource exploitation, waste pollution, and devastating working conditions are a few of the topics we will touch in this book. Like the coronavirus, they are invisible, and harm or destroy life. As if under a magnifying glass, these times reveal where our world is broken, where economic models fail and which of our societal and personal values and attitudes nurture life. This year also showed that we as individuals can influence the course of things at a very personal level through our behavior–which is applicable to sustainability, too.

The world will go on, no matter what we do. But I´m convinced we can make choices to direct it.

As a business person, I believe companies have a major role to play in sustainability. Over the past almost two decades, I have had the privilege of navigating many companies, in all types of industries, through the process of initiating and expanding their sustainability management and communications. I would like to make a few things I have learned accessible to a broader audience. I believe we need better and more focused management and communications when it comes to sustainability. We need a new propelling power originating from corporate action, even while sustainability´s importance is dwindling in the light of Covid-19´s challenges.

"DO YOU SPEAK SUSTAINABILITY?" invites you on a personal learning journey. Whether you are new to sustainability or have been involved in it for some time—each of us can learn more and act differently. To bridge the demands and necessities of newcomers and the knowledgeable alike is a demanding, self-imposed target. It stems from my didactical understanding of how we learn and come to action in sustainability. As complex and multi-faceted as sustainability is, it needs to be adapted to the specific context. We must continuously learn to increasingly recognize, evaluate, consider and apply sustainability aspects to new topics and arenas.

In this book, I strive to explain, contextualize and illustrate this with data from different companies and industries. I invite you to apply these lessons learned to your own situation and context. As a consultant, I love frameworks and benchmarking. Yet, there are limitations of generalized and simplified approaches. Navigation in sustainability implies empowering people to see and judge differently, broader, more integrated.

Although the spotlight is on my special field of expertise–corporations, companies are composed of people. This is why I call it a personal navigator for corporate action. Using this book as a personal navigator, can help to become aware of the different roles we play as individuals (as business leaders, friends, consumers, family people, citizens, to name a few) on this journey. I have highlighted our role as consumers in a dedicated chapter. My invitation is to read this book in the light of your personal contribution within your individual circle of impact. If it can inspire you to do three impactful things towards the change you want to see, then it was worth all the effort pulling it together.

This is also a personal journey in the narrower sense: As people, we all have values, experiences, goals, fears, failures, and hopes that shape and impact our paths. I share some insights from own personal learning journey with respect to my underlying values as a believer in the epilogue.

The year 2020 interrupted familiar routines. This can be an invitation to pause and change the way we live together and do business.

Now is the time to take a new path in this journey. I am excited to be able to accompany you on your way.

Sincerely Yours,

Hamburg, December 2020

BUSINESS IN THE 21ST CENTURY

HOW TO USE THIS BOOK

This is a reference book about sustainability that focuses on corporate action.

Chapters 1-3 elaborate on some of the general concepts and dynamics that are valuable as well as necessary for understanding sustainability.

Chapter 4 focuses on managing sustainability.

The four subsequent chapters focus on specific topics of interest; namely, communications (chapter 5), innovation (chapter 6), people (chapter 7), and community engagement (chapter 8). I suggest first reading chapters 1-4 unless you are a practitioner already familiar with the key topics of managing sustainability.

Chapter 9 dives into selected industry dynamics when interfacing with consumers.

Chapter 10 highlights how to be on the sustainability journey as consumers.

The final pages provide a few jumping-off points for continuing your personal journey.

HIGHLIGHTED

are some concepts and key terms frequently used in sustainability. They are not just defined but also put into context.

TIPS

in each chapter provide further inspiration or encourage you to deep-dive further into a topic. They include carefully selected articles, videos, elaborated reports, and links, as well as examples for further illustration of the topics discussed in the passages.

 **SPOTLIGHTS**

can be found in some chapters explaining and illustrating important topics and providing detailed data.

 **TOOLBOXES**

are provided starting in chapter 4. They contain checklists, how-to's and recommendations on how to handle specific tasks.

APPLICATIONS

are found at the end of each chapter and are an invitation to reflect on the implications for your own context. Use these when suitable and allow yourself some time to work through them. They also include some quizzes for a fun break.

This book is not meant to provide quick fixes and will not likely answer all of your questions. Its intention is to help enrich your knowledge and empower you to apply sustainability aspects within your specific context. In "How to get in touch" (see page 352), I provide some ways you can get in touch for guidance during your own individual journey.

Note:

Data and examples are from a wide range of industries and countries but are naturally focused on the contexts I am most familiar with. If you have any other suggestions or know of some good examples or ideas worth spreading, please let me know!

This data has been researched and edited carefully and consciously, to the best of my knowledge and belief. If you have any comments or questions regarding data, please feel free to contact me.

E-mail: hello@sustainavigator.com

www.sustainavigator.com

CONTENT OVERVIEW

CHAPTER OVERVIEW

I remember the children's church camp where I held
a workshop on economics, discussing with the kids on
a simple and practical scale about the willingness to pay
(for a chocolate bar), currency exchange rates
(while traveling abroad) and global supply chains
(to produce a T-shirt). My passion for imparting knowledge
met with my love for economics and business.
All of the few kids attending—most had chosen football,
handicrafts and makeup classes instead—had their touching
"aha moments" about economic insights.

It is two of today's great divides: Our world is
dominated by economic interdependencies, but we
understand and care little about them. And the current
narrative of business does not propel sustainability.
How in this setting, can companies take a new stance
as significant actors?

TIME FOR RESPON-SIBLE BUSINESS

Businesses should provide safe, good-value products and services at a fair price. Right? But it's not that simple anymore. Businesses also provide jobs, pay taxes and drive innovation. What else is expected of businesses today? Before we start looking closer at the role companies play in today's world, we need to understand why their role has become a matter of concern in the first place.

COMPANIES AS PART OF SOCIETY

At the beginning of my university lectures on sustainability and corporate responsibility, I would show the students a simple slide about companies as part of society. This would launch us into a discussion about the principles of business and economics and people's perceptions and their experiences with companies. Such a broad view on inputs, outputs and impacts was a good starting point to explore the responsibility companies have in today's world.

Business as part of society

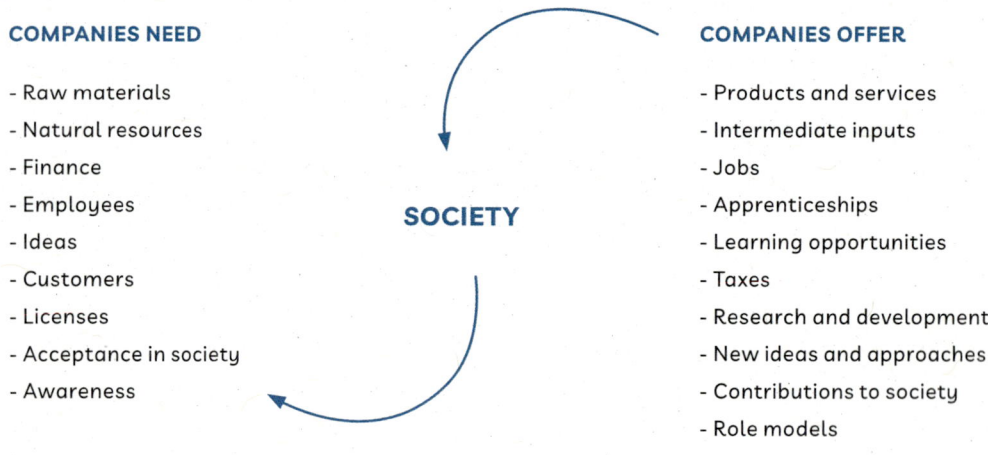

COMPANIES NEED

- Raw materials
- Natural resources
- Finance
- Employees
- Ideas
- Customers
- Licenses
- Acceptance in society
- Awareness

SOCIETY

COMPANIES OFFER

- Products and services
- Intermediate inputs
- Jobs
- Apprenticeships
- Learning opportunities
- Taxes
- Research and development
- New ideas and approaches
- Contributions to society
- Role models

Own graph

Companies traditionally have strong and complex interfaces with many spheres of society. Value chains of companies are often deeply wired to the local and the global economies. Business is a part of society by its nature.

TIP

An insightful discussion of business leaders looks at the detachment of companies and their role in society.

tinyurl.com/y3qb2d9o

Yet, they are often not fully perceived and appreciated in this core position in society. There are many reasons for this, some of which we will explore throughout this book. For now it is relevant to consider the fact that companies have been detached somewhat from society for a long time and have acted primarily within their own business ecosystem. This occurred at a time with a generally low level of understanding of economic interdependencies within our society—which has further supported business' alienation from society.

Business today is much more willing to take on the role (again) of a significant actor in society. Yet, these efforts have been met by an overall atmosphere of society's distrust of companies and institutions in general, as multiple statistics show. This situation will likely remain unchanged for some time and form the fundamental environment in which many companies operate their business and apply their sustainability approaches.

MISSING THE BASELINE

Most people would probably expect companies and their leaders to demonstrate a minimum of at least two things: lawfulness and some basic decency. Most companies do meet this expectation, and there are countless managers and company owners that also do a great job of leading their companies.

Corporate scandals are the exception and not the norm. Yet, they present a strong public image of influential people acting out of greed, selfishness and recklessness. Systems, both internally and externally within the company's operating environment— includes regulation and control— fail to prevent this. People fail to stand up. An example of this at a broader scale occurred with the 2008 financial crisis, which severely deteriorated public trust. Government bail-outs of large corporations also created further distortions and divide. In this short documentary on the biggest corporate scandals, the ugly side of business is revealed with concrete examples. ⌁ tinyurl.com/y2pxmc6l

Aside from the big scandals that make it to the media, cases of fraud, bribery, corruption, sexual immorality and harassment, tax evasion, money laundering, market manipulation, non-transparent financial transactions, lobbying against stronger regulation, and double standards within companies have been and still are entirely too common.

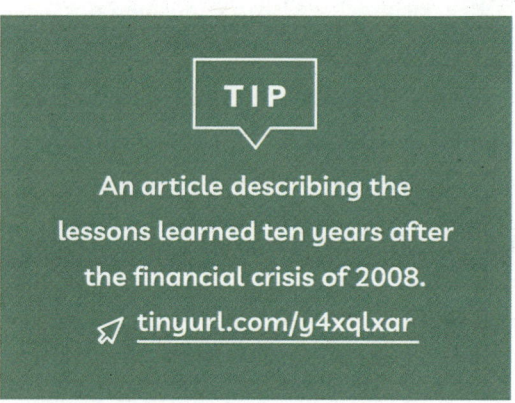

TIP

An article describing the lessons learned ten years after the financial crisis of 2008.
⌁ tinyurl.com/y4xqlxar

Over the years, they have led to the development of extensive corporate governance and compliance measures and structures. Corporate governance deals with internal control and encompasses the work of the board, as well as external relationships with shareholders and other stakeholders (see pages 74ff.). Compliance is about the processes through which companies try to ensure that employees and other relevant actors related to the business (such as suppliers) act in accordance with the law and other applicable norms. Often, risk management is associated with compliance management. Corporate governance and compliance constitute a formal, primarily legal, base for doing business. In this book, it will be addressed only in relation to sustainability, such as in supply chains.

In many cases in the past, lawfulness and decency, in terms of basic ethical norms and values, were missing. With attitudes often lacking responsibility, trust into companies, economic systems and business leaders as role models is not self-evident. Without overstating the point, business leaders carry a heavy legacy of public distrust.

It's part of the narrative that responsibility is not instinctively at the core of business leaders' concerns. That corporations try to push the boundaries as much as possible to avoid responsibility where mandatory regulations do not exist, such as in social, environmental and ethical areas. And that there are few topics companies will address unless they are forced to or only when it helps their reputations. In other words, they care about little else than money, power and their own interests. That's the (somewhat sobering) basis from which we address sustainability—an urgent, complex and delicate topic.

TIP

An article on the future of Corporate Governance.
tinyurl.com/karmjb9

A 2018 study on the state of compliance.
tinyurl.com/yxzlunza

WHAT IS BUSINESS ABOUT?

Let's take one step back and look at the original objective of business. A company, according to the Merriam-Webster Dictionary, is an association of persons for carrying on a commercial or industrial enterprise. A broader scope unfolds when looking at the definition of the term "business."

(selection and own italics):[3]
– a commercial or industrial *enterprise*
– *serious activity requiring time and effort* and usually the avoidance of distractions
– an immediate *task or objective (mission)*
– a usually commercial or mercantile *activity* engaged in as a *means of livelihood*
– dealings or transactions especially of an *economic nature,* in the understanding of *patronage*
– a particular field of *endeavor*
– *creation*

The definition reveals the nature of how business is meant to be. Business provides goods and services to people and asks a price for this in return. It is a mandate—resources are entrusted to a company to create value, requiring dedication and resulting in providing for oneself as well as contributing to a greater cause. Companies are on a journey of discovery, entitled to provide, serve, create, contribute and nurture life. It is a great adventure! I believe that this perspective draws on a broader vision of business which may have been lost sight of today. To rediscover and reconsider this core nature and broad vision of business could be a good start in the countless discussions about purpose that currently take place in many companies (see page 218).

THE BUSINESS OF BUSINESS

Economist Milton Friedman stated in a 1970 article: *"The business of business is to make profits"* or, to put it another way, *"The business of business is business"*. Friedman has been cited repeatedly over the past decades and has evolved into somewhat of an antagonist to sustainability and corporate responsibility. His perspective contrasts with an emerging culture of companies that care for people and nature instead of only profit.

My preference is to argue that if businesses were conducted in the first place as described above, profitmaking would be not a problem but a wise way of stewarding limited resources.

As already shown, when businesses interact within society, it's hard to draw a strict line of the business of business anyway. Indeed, some companies over the past two decades were even confronted with having their license to operate at stake: Is there still a "reason for being"? Is the company socially accepted? This was less about the public criticism of certain practices and more about the overall acceptance and fundamental questioning of the business model and corporate reputation.

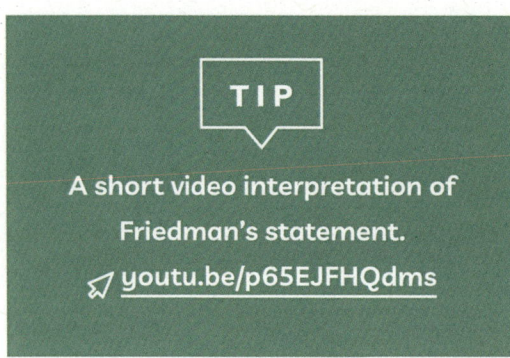

TIP

A short video interpretation of Friedman's statement.

⊲ youtu.be/p65EJFHQdms

In the case of tobacco company Philipp Morris, this resulted in the public regulation of marketing, as well as a bad image for the whole industry. The company has started the long road to transformation (see page 169). Pharmaceutical companies were caught in the public crossfire in the early 2000s. Numerous cases of unethical marketing had damaged their reputations. The business model of pharma companies impeded access to medicine for poor people around the world. The license to operate became a public issue and forced the pharmaceutical industry to undertake a tremendous transformation in access to medicine (see page 217). Energy giants and oil and gas companies have also seen their license to operate deteriorate over the past decade, fueled by the debate on climate change. BP CEO Bernard Looney in a recent public speech, described the process of detachment, the loss of license to operate and the necessity to regain it (see page 171).

His speech is a didactic play for any manager on a transformational journey: tinyurl.com/y2esezzq

Regaining the license to operate is a long and difficult road to regaining trust through open communication and fundamental change.

TIP

For deep-diving into Milton Friedman: An article with interesting transaction-based reflections.
tinyurl.com/y4yysks6

Group discussion materials featuring Friedman's arguments.
tinyurl.com/y6bb4tcf

1.02 ——— POWER MEANS RESPONSIBILITY

This book looks primarily at large companies that operate in a globalized world. They are large in economic terms as expressed by conventional measures. How does this translate into power, meaning importance? And which role does responsibility play?

AN EMPLOYMENT ENGINE

An important public contribution of companies is often taken for granted but has significant value to people and societies: Corporations are major employers. Fortune magazine's 2019 list of the 10 largest employers showed that they accounted for 9.2 million jobs.[4] Among the biggest employers in the world are China Post Group, US Postal Service, China National Petroleum, Volkswagen, Amazon, and Walmart. In spring 2020, Amazon alone hired 75,000 new employees in the US to meet the rising demand during the Covid-19 crisis.[5] Other huge, non-listed employers include the Tata Group, based in India and operating in 100 countries with 720,000 employees[6], German retailer Schwarz Group with 429,000 employees[7] or Bosch Group, generating in 2019 with 400,000 employees worldwide revenues of 77.7 bn euro[8]. As we will see later, the role of "employer" significantly impacts people and nations and is today a cornerstone of corporate responsibility. Employment is also a major factor in terms of contributing to the economic welfare of developing nations, as shown in a recent study of the World Benchmarking Alliance (see page 247).

Yet, when looking at employment, it is small companies that provide the largest share of jobs. According to the US Bureau of Labor Statistics, in 2019, a total of 74.6% of all companies in the United States had between 0 to 9 employees, and only 0.22% had more than 1,000 employees.[9] In Germany, 2.1 million out of 3.48 million businesses have between 0 and 9 employees (60%) and 300,000 between 10 to 49 employees (9%). Only 179 companies have more than 5,000 employees and 101 companies more than 10,000 employees.[10] Smaller companies also support the renowned German apprenticeship system. Most young people in Germany learn their profession at small companies: 29% of all apprenticeships are in companies with more than 250 employees and 71% in smaller companies, of which 16% are in firms with less than ten employees.[11] So, all companies small and large are contributors to our societies in providing employment, as well as in other ways of value creation. This book, however, focuses mainly on larger corporations.

SIZE MATTERS?

If Toyota Motors were a country, it would be number 23 in size—making it bigger than India or Switzerland, states the World Economic Forum (WEF) in a clip. tinyurl.com/yxh466jk

In the associated article, the WEF cites a ranking of the NGO Global Justice Now that compares government and corporate revenues. According to the article, of the world's top 100 global economic entities, 69 of them are corporations, leaving room on the list for just 31 countries. The countries in the top 10 are accompanied only by Walmart, which is the world's biggest company by revenue.[12]

TIP

An Economist magazine special report on the power of companies (2016).
tinyurl.com/y29hfnwm

Rankings by size are typical in the corporate world, as strength and significance are part of the economic narrative associated with size. The approaches are numerous. In Fortune magazine's ranking of the Global 500 by revenue, Walmart is joined by the two Chinese businesses petroleum and chemical company Sinopec and electric utility company State Grid. Other oil and gas and energy companies join the top ten as do two car manufacturers—Toyota as the top-ranked in terms of revenue and Volkswagen as number 7.[4] Measured by market capitalization, currency exchange platform Forex lists oil and gas producer Saudi Aramco as the most valuable company in 2020, followed by Microsoft, Apple, Amazon, Alphabet (parent company of Google) and Facebook.[13] Forbes Magazine publishes a broader ranking of the biggest companies worldwide called Global 2000, which is based on a combination of revenue, profits, assets and market value. In 2020, the top 3 companies were banks: Industrial & Commercial Bank of China, China Construction Bank and JP Morgan Chase.[14]

All of these rankings are based on size and a narrow perspective of economic output. It is a prevalent approach in today's business world that size defines value. The narrow view is on monetary value. The value that companies create for society, how they exert patronage and contribute to livelihood, as in the broad vision described above, is not in the focus of typical economic metrics.

SIGNIFICANT REACH AND IMPACT

As companies engage in wide-spanning activities, they influence many regions in the world, either directly through their operations or sales areas. The global reach of a few selected companies is shown in the table below.

Company	Number of countries where jobs are located (production/office sites)	Number of countries where products/services are sold
Unilever[15]	Ca. 100	More than 190
Amazon[16]	30	130
Volkswagen[17]	31 (123 production/office sites)	153
Novartis[18]	Over 50	Ca. 155
BASF[19]	Over 90 (367 production/office sites)	More than 190

Own presentation based on company data

Who benefits from the companies being in those countries? This quantitative perspective does not yet reveal whether or not companies leave an overall positive impact at those locations with or beyond their products and services. Yet, this perspective shows that big companies have an influence on a systemic level through their wide reach.

It also reveals the reach of companies in concrete numbers. Big companies reach many people, as the following examples show:

– Unilever's products are used every day by 2.5 billion people around the world.[15]

– Nestlé works with almost 165,000 direct suppliers and 695,000 individual farmers worldwide.[20]

– McDonald's serves 69 million customers each day.[21]

– HSBC bank operates in 64 countries serving more than
40 million customers.[22]

– Coca-Cola sells 117 billion non-refillable plastic bottles each year and a
total of over 225 billion units of its beverages, which is equivalent to 30
units per person worldwide.[23]

Whether or not a company leaves behind a positive or negative impact on
people and nature is not always easy to measure. Yet it is important to
notice: Size matters. Through the extent of their reach, companies exert
influence, which is why it is so important how they shape it. How
companies take on the responsibility resulting from their significant resources and impacts from their business will be a focus of this book.

There is a broader call to action for engagement in global
challenges. The Secretary General of the United Nations, H.E. António
Guterres, stated the role companies play in sustainability:
*"We have to mobilize the private sector, it is 75% of the global GDP. Moving
forward, collaboration with business—and the key CEOs in the world—is
crucial when it comes to fighting climate change; but also, to meet sustainable
development goals, eradicate all extreme poverty by 2030, and we're not on
track on this."*[24]

A SNAPSHOT OF NEGATIVE IMPACTS

Throughout this book, we look at the impacts of companies in different
industries. At this point, I would like to mention three examples that point
out the significant negative impacts associated with big business.

A 2019 audit conducted on plastic trash in 51 countries around the world
concluded that the top 5 plastic polluters are the global consumer
goods companies Coca-Cola, Nestlé, PepsiCo, Mondelēz and Unilever.
⤢ tinyurl.com/y2j264nx

A statistic on global emissions showed that just 100 companies have been
responsible for 71% of the global emissions since 1988. The Carbon Majors
Report of the Carbon Disclosure Project (CDP, see page 86) linked

emissions to the corporate level and concluded that a relatively small set of fossil fuel producers and their respective investors might hold the key to systemic change on carbon emissions. Over half of the global industrial emissions produced since 1988 can be traced to just 25 corporate and state producers; among them companies such as ExxonMobil, Shell, BP, Chevron, Peabody, Total, and BHP. Producers also included state-owned entities Saudi Aramco, Gazprom, National Iranian Oil, Coal India, Pemex and CNPC, as well as the Chinese coal industry.[25] ⟁ tinyurl.com/y88jplex

A similar finding in another area came from a 2019 research study conducted by the Stockholm Resilience Centre, cited in a publication of the World Benchmarking Alliance: *"a handful of transnational corporations were responsible for most of the impacts across agriculture, forestry, seafood, cement, minerals and fossil energy."* Huge actors are assumed to have disproportionate effects on the structure and function of the system in which they operate.[26] One of the reasons why this book repeatedly addresses huge corporations is for just this reason: their actions and transformation undoubtedly matter significantly.

1.03 ——————— OUR ECONOMIC NARRATIVE

In the next chapters, we will take a deep-dive into sustainability. Let's therefore take a quick look at the economic narrative into which it is embedded. Business is based on assumptions and mechanisms and operates within a certain worldview. I would like to comment on three aspects of our narrative.

FREE MARKETS

Most of today's nations operate in free market systems in which the protection of personal property is a major foundation. A fundamental value is that individuals are able to act independently. The underlying economic theory—known as homo economicus—describes humans as rational and self-interested beings. They take decisions economically,

GLOBALIZATION

The increasing level of the economy's global integration has accelerated and facilitated global business.

Below I present some selected data on the state and trajectory of globalization published by global logistics specialist DHL. The DHL Global Connectedness Index measures globalization based on the flows of trade, capital, information, and people. The company's reports and its website provide interesting and, in some cases, surprising insights about globalization. ✈ tinyurl.com/y5g3rk6b 2020 update: ✈ tinyurl.com/y5u4qop6

Selected insights on globalization from the DHL Global Connectedness Index 2018 and 2019 Updates.[27]

– The Global Connectedness Index shows that the world's level of connectedness has risen significantly since 2001. All categories—capital flows, trade flows, people flows, and above all, flows of information—have increased.

– Exports as a percentage of world GDP were 25% higher in 2018 than in 2000, more than twice as high as in 1970, and almost six times higher than in 1945.

– The world's most globally connected countries in 2017 were the Netherlands, Singapore, Switzerland, Belgium, the United Arab Emirates, Ireland, Luxembourg, Denmark, the United Kingdom, and Germany. The list of the top 60 countries include representatives from all geographic regions.

– While the world is more connected now than at almost any time in the past, international flows are far smaller than most people would presume. Globalization is significant. But most business still takes place within rather than across national borders.

– The combined output of all multinational firms outside of their home countries added up to only 9% of global economic output in 2017; just 2% of all employees around the world worked in the international operations of multinational firms.

– Most companies are still domestic. Less than 0.1% of all firms have foreign operations, and about 1% of companies export. On average, small firms are much less international than large ones, and most companies are small. But even among the Fortune Global 500, the world's largest firms by revenue, domestic sales still exceed international sales.

Check also out the World Bank with a broad range of global data and stories as well as an interesting look at world economic dynamics.
tinyurl.com/y4s6u8t8

and the self-interests of different agents (individuals, firms) lead to an equilibrium between supply and demand through market mechanisms. Prices and quantities for goods organize supply and demand in a market economy. If demand is high or availability is scarce, prices rise and, subsequently, supply catches up. The market essentially regulates itself through transactions in the markets. This is basically the free market narrative.

Several assumptions, such as information transparency and rational behavior, have been discussed for decades by economists. The point I would like to stress with regard to sustainability in this part of our economic narrative is the following: Most markets do not actually function very freely. They are grounded in laws and regulations of all types in terms of product safety, data security, environmental protection, and steering mechanisms such as international standardization schemes and measures to protect vulnerable groups not covered by free market mechanisms, just to name a few. Covid-19 strongly revealed markets' need for intervention through policymaking. Most sustainability topics now and in the future will need some kind of steering and guidance as this cannot be provided by free market mechanisms alone.

UNLIMITED NATURAL RESOURCES

Prevalent economic theory is based on the underlying assumption that there are no limits to natural resources. Major elements of production such as labor, capital, and land are seen as interchangeable. That is to say that resource owners will seek the most profitable use of their resources which, under (perfect) competition, results in an equilibrium. There are no boundaries assumed in our economic systems. The principle of economic efficiency provides the balancing effect: productivity increases and technological leaps lead to progress and, for example, decouple growth from material use. As two economists summarize it: *"With few exceptions, economics as a discipline has been dominated by a perception of living in an unlimited world, where resource and pollution problems in one area were solved by moving resources or people to other parts. The very hint of any global limitation ... was met with disbelief and rejection by businesses and most economists. However, this conclusion was mostly based on false premises."*[28]

One missing link in the narrative lies in the fact that most natural resources (oceans, air, landscapes, species, etc.) do not have a price tag but are basically treated as freely available. Often, prices for raw materials and some products do not include the full costs incurred by people and nature. Market mechanisms fail when the originator doesn't have to bear the full costs. Such externalities frequently occur, for example, in relation to pollution or precarious working conditions in supply chains. Also, efficiency often leads to increased pressure on natural resources due to rebound effects (see page 60). There is still a question as to how this can be addressed. I assume, in line with other experts, that this will frequently turn back into a policy issue.

TIP

A worthwhile video from the perspective of MIT economist Andrew McAfee, author of the book "More from less."
He has an optimistic but inconclusive view of the ability of market mechanisms to solve today's challenges.

tinyurl.com/w2mm7af

THE KEY: PERPETUAL GROWTH

The basic narrative of our economies is the goal of growth and the continuous desire for more material wealth—not only from countries and companies but also consumers. Growth must continuously be nurtured. The central theme of the narrative, in short: continuous consumption keeps the economy going, saves jobs, creates shareholder returns and tax income, and GDP growth raises the wealth of a country. At a company level, growth in returns and profits, as well as expectations for future growth, will be measured on the stock markets in the near-term and ongoing period, primarily in economic terms. At an individual level, people have unlimited desires that can and need to be met by ever new products and services that they want to buy or are triggered to ask for.

Much has been said and written about the growth debate—let me highlight three points in a nutshell under the assumption that our prevalent growth narrative cements an unsustainable path. On the one hand, evidence reveals a close correlation between economic growth and the rising rate of carbon dioxide emissions, i.e., growth nurtures climate change. Secondly, growth does not imply wellbeing—the just distribution of income for example or the consideration of other factors such as health. Thirdly, the continuous consumption cycle under today's conditions leads to high social and environmental costs, as we will see. The narrative that growth is an end in itself and provides value starts to show some cracks.

Scientists who conducted a major assessment on biodiversity stated in a 2019-released report: *"We need to change our narratives. Both our individual narratives that associate wasteful consumption with quality of life and with status, and the narratives of the economic systems that still consider that environmental degradation and social inequality are inevitable outcomes of economic growth. Economic growth is a means and not an end. We need to look for the quality of life of the planet."*[29]

TIP

An OECD initiative to measure the wellbeing of nations more broadly.
✈ tinyurl.com/y2eotf22

IS THERE A NEW NARRATIVE?

This book will focus on business transformation towards more sustainability which continues to basically take place within the current economic narrative. At this point, I would like to emphasize, on a general level, how companies can expand on their role in society going forward.

A NEW ROLE FOR BUSINESS

"... (There is) an imbalance between the value that a company creates for itself and the value it creates for others (which) will lead to an unsustainable business. ... The potential and need are huge and the time to collaborate on a new path is now. As the leaders of today, we have an obligation to do business better." stated Stephen Badger, Director and former Chairman of the Board of Mars in 2020.[30]

Indeed, there is an increasing understanding of a new role for companies, as well reflected in the above citation. Things have to change for big businesses. To get there entails a huge transformational journey that we will look at in the coming chapters. In the light of the last passage, a provocative question arises: What would it mean to draw an alternative to the growth narrative for a major multinational business such as Mars? Which boundaries could be set? It is an interesting and important question but not possible to address within the scope of this book. However, I am convinced that during the next decade, questions such as these about big business will increasingly be discussed and alternative approaches to growth will emerge. But when it comes to big business, we are not at that point yet. This is why I would like to focus on the current stage and pave the way for the transformational journey.

We will see other initiatives, too, such as compensating for business activities and their effects. Microsoft, for example, recently announced that it would offset all of the company's carbon emissions since its establishment in 1975. By 2050, the company will have removed from the environment all of the carbon it has directly emitted or produced by electricity consumption since its foundation. Microsoft will basically erase its entire carbon footprint. While carbon offset is critically discussed, done correctly, it can be a relevant and valuable contributor to fighting climate change (see page 342).

↗ youtu.be/LeQxTI-s48A   ↗ news.microsoft.com/climate

CREATING VALUE FOR SOCIETY

An approach we already see on a growing basis is companies' focus on the value they create. Businesses specify their role in society and, they do so to externally communicate the value they contribute. This increasing transparency also allows for comparability and keeping companies accountable.

A good example—also with respect to visualization—is Evonik, an internationally leading specialty chemical company based in Germany. Evonik publishes an extended picture of the value chain, mentioning and quantifying its contributions and impacts using a wide variety of inputs transformed into outputs and value (see pages 36-37).

Similarly, Novartis, the aforementioned global pharmaceutical company, measured and disclosed how much value it created through its business operations in 2019. It shows that shareholder returns represent just one component among many others.

Novartis in society

THE VALUE WE CREATE

JOBS	SHAREHOLDER RETURNS	TAXES PAID	IMPROVED HEALTH AND WELL-BEING	ACCESS TO MEDICINE AND HEALTHCARE
1.3 m	**22.3 %**	**1.9 bn**	**67 bn**	**799 m**
Indirect, induced (2018)	2019 total shareholder return (USD), including Alcon spin-off	(USD)	Social impact (USD, 2018), based on estimated value of health benefits to patients	Patients reached with Novartis medicines
108 775	**6.6 bn**			**10 m**
Own operations	Total dividends paid (USD)			People reached through training and health education programs

Source: Novartis in Society. ESG Report 2019. https://tinyurl.com/yys3avfx. Accessed December 7, 2020. With permission of Novartis.

This transparency of sustainability is increasingly expected of companies. Publishing the value chain and specific performance indicators is a disclosure requirement in reporting. It is also, in my experience, a significant step in the transformational process. Companies seek to become increasingly self-aware of themselves and their role. They discover a new way of seeing who they are, what they are there for and what makes them unique. Under the spotlight of sustainability and increased expectations, they re-define their role and strive to be perceived as an active, relevant player in society that contributes to a greater cause.

Frequently, it is assumed that companies are expected to do so. I often see this as a call to action rather than an (enforceable) expectation. Jay Coen Gilbert, an entrepreneur, describes his view in a Forbes article:

"Fast growth is one measure of a business' success, but it is no longer enough. The world needs businesses that are both high growth and high impact. The world needs businesses that make "yes/and" the norm, and leave the old thinking of "either/or" behind. The world needs social unicorns who, instead of aiming to hit $1 billion, reach to touch one billion lives ... Businesses that compete not only to be best in the world but also to be best for the world. As consumers, talent and investors increasingly seek values-aligned businesses to buy from, work for and invest in, entrepreneurs are seeking to create the greatest positive impact not just to be nice, but to be more successful." [31]

THE TIME IS NOW

In 2019, 11,000 scientists from 150 countries declared the climate emergency. They pledged that *"our goals need to shift from GDP growth and the pursuit of affluence toward sustaining ecosystems and improving human wellbeing by prioritizing basic needs and reducing inequality."* [32]

The decade 2020-2030 is seen by many experts as a decisive one for switching course. Not only Microsoft referred to it, but also other big businesses are calling for "a decade to deliver" (see page 136). The United Nations set the Global Agenda until 2030 (see page 52).

Resources and value contributed in 2019

Our resources »

Society

32,423	employees
approx. 29,000	customers
approx. 30,000	suppliers

The environment

63.49 PJ	energy inputs
534 million m³	water intake

Employees

101	nationalities
25.7%	female employees (Evonik Group)
26.1%/24.1%	female managers at the 1st/2nd management level below the executive board

Financials

€6,435 million	property, plant and equipment
€842 million	capital expenditures

Knowledge

approx. 24,000	patents
approx. 2,560	R&D employees
€428 million	R&D expenses

Production

€9.4 billion	procurement volume
9.24 million metric tons	raw material inputs
7.9%	renewable raw materials

[a] Outside the scope of the limited assurance review by PwC.

[b] For further water data, see chart **C21**.

[c] Product sales are patent-driven if there is at least one relevant global patent.

Our busi

R&D as a driver for resource-saving products

>100 production sites worldwide

€**2.15** billion adj. EBITDA

Sustainability as a growth driver for our **markets and customers**

approx.**29,000** customers

SDGs of relev

...ss »

approx. **30,000** suppliers

We assume responsibility for **production and the supply chain**

€13.1 billion sales

>4,000 products

Recycling/ disposal

...ustomers   End-customers

for Evonik:

Value contributed »

Society

approx. **€2.5 billion**	wages and salaries
€3.7 million	donations and sponsorship [a]

The environment

5.5 million metric tons	CO_2 emissions (scope 1 and 2)
12 million m³	water consumption [b]

Employees

0.9%	early employee turnover
€76 million	vocational and advanced training
1.18	accident frequency

Financials

4.1%	dividend yield
€180 million	income taxes
€64 million	interest and other taxes

Knowledge

approx. **225**	new patents and patent applications
13%	sales with products and applications developed in the past five years
47%	patent-driven sales (based on total sales) [c]

Production

1.10	accident frequency
97.1%	of all sites are certified in conformance with ISO 14.001/9001
9.16 million metric tons	output

Evonik Sustainability Report 2019. tinyurl.com/y6m7vnfg. Accessed November 28, 2020. With permission of Evonik.

TIP

Rankings of companies that add value to society are published regulary, as the "Best for the world honorees".

↗ tinyurl.com/y3egqpgk

Then, the year 2020 came, and Covid-19 hit. I started to write this book right before the crisis began. And while the year 2020 was dominated by Covid-19 and the economic crisis, the call to action has not ceased. As mentioned in the introduction, when US author Stewart Brand was asked what the corona crisis could teach us about climate change, he pointed out that although climate crisis is a much bigger and deadlier issue, it will not see the same immediate reaction because of its creeping approach. He made an illustrative point: *"Civilization is stuck in puberty. We think short-sighted, egoistically and live in the misconception to be immortal. The task of this century will be that we work us out of this chaotic phase that is marked by illusions and that we overcome the pathologically short attention span of our civilization."*[2]

The previously cited Mars executive also referred to the current situation: *"The world is at a crossroads. The COVID-19 pandemic has shed a glaring light on how much is broken in our social and economic models. It is imperative that we consider our present state and the future we want. Our opportunity is to define recovery not as returning to what was, but rather to choose a more sustainable and inclusive road forward to help us all thrive."*[30]

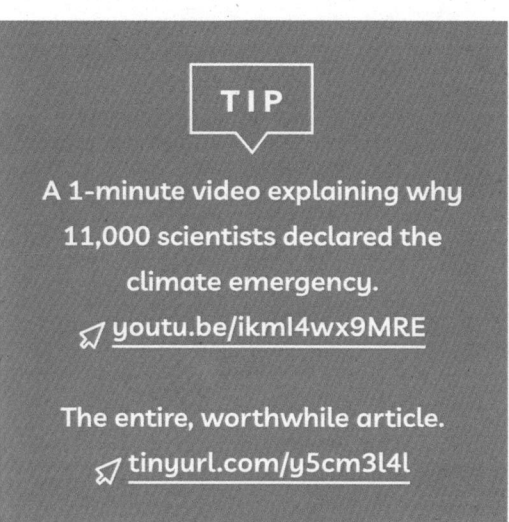

TIP

A 1-minute video explaining why 11,000 scientists declared the climate emergency.

↗ youtu.be/ikml4wx9MRE

The entire, worthwhile article.

↗ tinyurl.com/y5cm3l4l

We will need a new narrative for companies and executives that marks a major transition. One that will involve creating value over economic growth, providing solutions for the world's problems and consequently phasing out unsustainable business practices. We are now at the very early stage of making this happen.

This transformation could fundamentally change our businesses and economies over the next decades. What's the direction? Let me illustrate it with an inspirational challenge for a product and industry we are familiar with. The call to action stems from a Greenpeace report, challenging businesses to catalyze the innovational spirit and creativity inherent to them.

"The smartphone is perhaps one of the best examples of human ingenuity of all time. However, the current production model is not one we would be proud to pass on to our grandchildren. … We are challenging all electronics manufacturers to imagine a new way—a business model that in 10 years' time will be unrecognizable compared to the current wasteful and harmful system. Imagine if technology was our strongest tool for creating a healthy, vibrant and thriving planet. Imagine if together we could harness technological innovation to help us overcome the Earth's biggest challenges by sharing ideas and solutions across the world. As IT companies have shown again and again, technology and creativity can be used as powerful forces to disrupt outdated business models. Leading IT companies can become the greatest advocates for a closed-loop production model and a renewably powered future. The brightest designers can create toxic-free gadgets to last, be repaired, and ultimately transformed into something new. It's time for the industry to adopt meaningful innovation—a slow, clean, closed-loop production model, powered by renewable energy. Who is going to be the first to take up the challenge?" [33]

APPLICATION

1. ——— Check out the International Business
Knowledge Quiz. ⟋ tinyurl.com/y6nhjlao

2. ——— Take a quiz on globalization with the informative
DHL Global Connectedness Index. Scroll down to
"Test your knowledge". ⟋ tinyurl.com/y5g3rk6b

3. ——— What is your view on how trust in companies forms a
basic condition for operating in sustainability?
How can you measure and improve trust for your
specific context?

4. ———— What is the economic narrative you operate in?
The following questions can help you to identify it:

– Which of the arguments in the passage on
economic narratives resonate?
– What keeps your company going?
– What keeps our economy going?
– Which market mechanisms work well,
and where do they fail?
– Where should more regulation be in place?
– Which motivations drive my consumption?

5. ———— Is it imaginable for your business to think of a
paradigm other than perpetual growth? How
would it look? What would it imply and require?

WHAT IT'S ALL ABOUT

For Greta, it's clear. It's all about climate change, her future, and the future of her peers. Fair enough. But is sustainability only about climate change?

"Sustainability" is used often and with little accuracy, which is why it is increasingly in need of a more concrete and precise definition. During countless projects and discussions with clients, teaching sessions and dinner conversations, I had the opportunity to translate sustainability into a specific setting: The topics that emerge differ greatly depending on whether it's a pharmaceutical, raw material, technology, dairy or other type of company. And as the example of my favorite drink, other than water (it's coffee), shows: In many cases, sustainability is also about individual preferences and considerations.

THREE SUSTAIN- ABILITY MARKERS

Let's start with some history. The earliest documented advocate of sustainability was the German accountant Hans Carl von Carlowitz, who introduced the term "sustainability" in a forestry context in his book Silvicultura oeconomica published in 1713. In the interest of conservation, he suggested using just the amount of wood that could be grown in a continuing, stable and sustained manner.[34]

Since then, the concept has evolved and multiplied, spreading into different spheres of business and everyday life. We will look at three specific markers that have shaped our actual understanding of sustainability over the past fifty years. All of them took place on the international stage, and all reflect something of the goals, the scope, and some of the biggest challenges and opportunities embedded in sustainability. The three specific incidents we will look at include a book released in 1972, an international conference in 1992 that laid the groundwork, and two significant agreements made in 2015.

CHALLENGING "BUSINESS AS USUAL"

In 1972, a book was released that still today is seen as a classic on economics and sustainability. "The Limits to Growth" was published by Donella H. Meadows, Dennis L. Meadows, Jørgen Randers, and William W. Behrens III, representing a team of 17 researchers from the Massachusetts Institute of Technology (MIT). The book consolidated in an easy-to-understand manner the results of a report commissioned by the Club of Rome, an international thinktank.

In essence, the researchers analyzed which impacts economic and population growth would have within a system of a finite supply of resources. Their computer simulation included five variables: population, food production, industrialization, pollution, and consumption of nonrenewable natural resources, all of which grew at the time of analysis. The authors intended to explore the sustainable feedback pattern when altering growth trends among the five variables under three scenarios. One of these was a "worst-case scenario", under which the planet would reach its boundaries within the next one hundred years. A "business as usual" attitude with no significant modification in human behavior would be characteristic for this scenario. The authors predicted and warned that out of the three modeled scenarios, societies would most likely choose this scenario.

TIP

The 1972 version is available in bookstores and to download.
tinyurl.com/qxohw4h
A 30-year update was published in 2004. In 2012, a 40-year forecast was published: "2052 A Global Forecast for the Next Forty Years."

And they were right. A 2014 article in the British newspaper "The Guardian" stated: *Research from the University of Melbourne has found the book's forecasts are accurate, 40 years on.*[35] The research showed that, based on data from the United Nations, the graphs almost precisely matched the worst-case scenario from 1972. "The Limits to Growth" report received high accolades and still today serves as a wakeup call. There were strong criticism and opposition over the years, including arguments that the methodology was flawed, the resources lacked the assumed limits, and technology's ability to fix problems was underestimated. The "business as usual" attitude, which basically reflects the economic narrative described, led to its obstruction by influential people and a loss of valuable time in its adaptation. A current advocate against the limits to growth thinking is the aforementioned economist McAffee, also from MIT (see page 31).

DRAWING A VISION

The United Nations Conference on Environment and Development, also called the Earth Summit, was held in Rio de Janeiro in June of 1992. The Earth Summit addressed diverse social, environmental and economic issues of global relevance and emphasized the global responsibility of intergovernmental bodies. An important achievement of the Summit was the agreement on the Climate Change Convention which would later lead to the Kyoto Protocol and the Paris Agreement in 2015. Other agreements and initiatives that were started at the conference addressed issues such as biological diversity, forest principles, the protection of indigenous people and combating desertification. Also, the "Agenda 21" was released, a non-binding action plan whose major objective is to launch local initiatives on sustainable development.

The basis for the Summit was "Our Common Future", a report published in 1987 by the United Nations. Known as the Brundtland Report, in recognition of former Norwegian Prime Minister Gro Harlem Brundtland, it placed sustainability firmly on the political agenda. The definition *"Sustainable development is development that meets the needs of the present without compromising the ability of future generations to meet their own needs."*[36] emphasizes justice between current and future generations. As a vision or overall guideline, it proposes a balance between economic development and social, as well as environmental aspects.

In many respects, the definition remains vague and generic, for example, in the question of how "needs" are defined. On a general level, it acknowledges boundaries but without specifying them. Despite all its limitations, this definition would become the basis for the developments in the years ahead and a guiding principle for the future. It was at this time that the World Wide Web had started its triumphal procession–which over time has also made the challenges and opportunities for business in today's globalized world more transparent.

TWO SIGNIFICANT AGREEMENTS

In late 2015, two international agreements came into existence: the Paris Agreement and the Sustainable Development Goals (SDGs). Both have had a significant impact on global efforts to advance sustainability.

In November/ December of 2015, climate change was widely accepted as a major challenge for humanity and for the first time internationally agreed upon under international law. At the United Nations Climate Change Conference (COP 21), the goal of limiting global warming to well below

2°C compared to pre-industrial levels was set with the Paris Agreement by the 196 participating countries. The parties agreed to pursue efforts to limit the temperature increase to 1.5°C which, according to scientists, would require zero emissions sometime between 2030 and 2050. The Paris Agreement entered into force on November 4, 2016, thirty days after the date the agreed threshold was achieved. At least 55 parties to the Convention, accounting in total for at least an estimated 55% of the total global greenhouse gas emissions, had submitted their instrument of ratification, acceptance, approval or accession.

TIP

The documentary "Guardians of the Earth" looks behind the curtains of the 2015 Paris negotiations. It puts stories and faces to the fight against climate change.

guardians.wfilm.de

Trailer: tinyurl.com/yxuqucac

The Paris Agreement requires all Parties to put forward their best efforts through nationally determined contributions (NDCs) and to strengthen these efforts in the years ahead. Each country that has ratified the Agreement is required to set a target for emissions reduction or limitation and regularly report on their emissions and implementation efforts. The target amount is voluntary, and no mechanisms exist for enforcing establishing it nor measures for when the target set is not achieved. Despite this, many protagonists join efforts to combat climate change, as documented by the progress of the Global Climate Action Agenda.

As the first and only country, the United States of America notified the Secretary-General on November 4, 2019 of its decision to withdraw from the Agreement, which took effect on November 4, 2020. An encouraging counter-movement of US citizens known as the "We Are Still In" initiative, a declaration and coalition formed by 3,800 politicians, businesses and other leaders from a broad spectrum of society. Mayors, governors and business leaders first began signing the We Are Still In declaration in June 2017 as a promise to world leaders that Americans would not retreat from the global pact to reduce emissions and would stem the causes of climate change. www.wearestillin.com

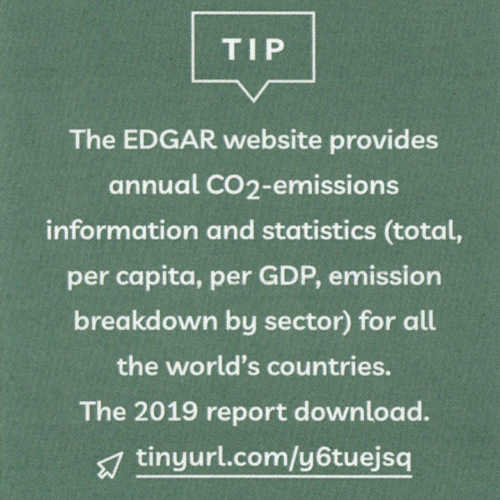

TIP

The EDGAR website provides annual CO_2-emissions information and statistics (total, per capita, per GDP, emission breakdown by sector) for all the world's countries. The 2019 report download.

⤻ tinyurl.com/y6tuejsq

Another business-driven initiative, "The Climate Pledge", was co-founded by Amazon in 2019. Under the pledge, companies commit to complying with the Paris Agreement by 2040, ten years ahead of time. Amazon's commitment, for instance, includes 100% renewable energy by 2025. Meanwhile, further signatories include, among others, Infosys, Mercedes Benz, Siemens, Schneider Electric and Verizon.

⤻ www.theclimatepledge.com
⤻ tinyurl.com/y2xskw33

The second significant agreement was adopted a few months ahead of the Paris conference. In September 2015, the 193 countries of the UN General Assembly agreed upon a global agenda for sustainable development and adopted the 17 Sustainable Development Goals (SDGs).

The SDGs succeed the eight Millennium Development Goals (MDGs) that were in place between 2000-2015. The focus of the previous development agenda was on the poorest populations in the world. The MDGs drew a positive balance. For example, from 1990 until 2015, the populations living in extreme poverty declined by more than half. The proportion of people undernourished in developing regions has declined by the same amount. The under-five mortality rate has declined by even more than half, and in 2015, maternal mortality was down 45%. Gains in the fight against HIV/AIDS, malaria and tuberculosis were another achievement in the MDG period.[40]

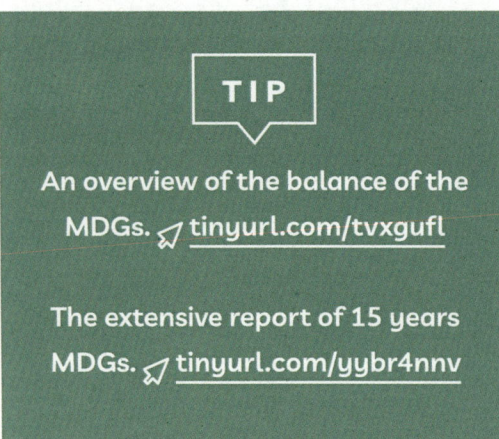

An overview of the balance of the MDGs. ⤻ tinyurl.com/tvxgufl

The extensive report of 15 years MDGs. ⤻ tinyurl.com/yybr4nnv

CLIMATE CHANGE

The Paris Agreement brings all nations together for the first time for a common cause to undertake ambitious efforts to combat climate change as a serious challenge. Carbon dioxide emissions have risen significantly since the start of the industrial revolution and over the past decades. Emissions have already risen from 22.7 gigatons in 1990 to 37.9 gigatons (2018). During this same period, the population has risen from 5.3 billion to more than 7.5 billion.

Today, most of the world's greenhouse gas emissions stem from a relatively small number of countries. China, the United States, and the European Union are the three largest emitters on an absolute basis, followed by India, Russia and Japan. Together these largest global CO_2 emitters account for 51% of the world's population, 65% of the global Gross Domestic Product, 80% of total global fossil fuel consumption and 67.5% of total global fossil fuel CO_2 emissions.[37]

Global carbon dioxide emissions by country in 2018

100%=37.9 Gt

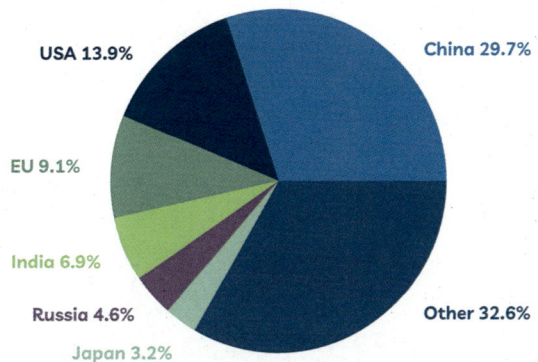

USA 13.9%

China 29.7%

EU 9.1%

India 6.9%

Russia 4.6%

Japan 3.2%

Other 32.6%

Own graph based on ibid[37]

Eleven other countries each contributed more than 1% to the share of total global emissions in 2018, namely Mexico, Brazil, Saudi Arabia, Canada, Turkey, Australia, South Africa, South Korea, Iran, Indonesia, and Germany (included above in the EU data). Per capita greenhouse gas emissions are highest in Palau, Curaçao, Qatar, Trinidad and Tobago, New Caledonia, Kuwait, United Arab Emirates and Bahrain. Among the largest emitters per capita are the United States and Russia; and among the EU countries, Estonia and Luxembourg. (own analysis, based on ibid)[37]

CO_2 accounts for 76% of total greenhouse gas emissions. Methane, primarily from agriculture, contributes 16% of greenhouse gas emissions and nitrous oxide, mostly from industry and agriculture, contributes 6% to global emissions. Figures are usually expressed in CO_2-equivalents (CO_2e/CO_2eq), converting other emissions into CO_2.

Worldwide, CO_2 per GDP has flattened and even slightly decreased since 1990. The contribution per sector varies at a global level. The power industry, for example, represents the largest share, which continues to grow rapidly, as does transport and other industrial (combustion for industrial manufacturing and fuel production). Other sectors, including industrial process emissions and agriculture and waste, have also risen significantly. Distribution and growth trends differ for each country, as shown by the specific developments for each country in the world contained in the EDGAR Report (see page 48).

In the EU28, emissions from fuel consumption have decreased in all sectors since 1990, except transport. In the context of the 2030 Energy and Climate framework, the European Union's current target is to reduce its greenhouse gas emissions by at least 40% compared to 1990 levels. By 2050, further reductions are needed if the European Union is to become climate-neutral, as outlined in the European Commission's vision "A Clean Planet for All."

Over the last few decades, global temperatures have risen sharply to approximately 0.7°C higher than the 1961-1990 baseline period. When going further back to 1850, temperatures at that time were an additional 0.4°C colder than during the baseline period. All in all, this

Development of fossil CO$_2$ emissions by sector

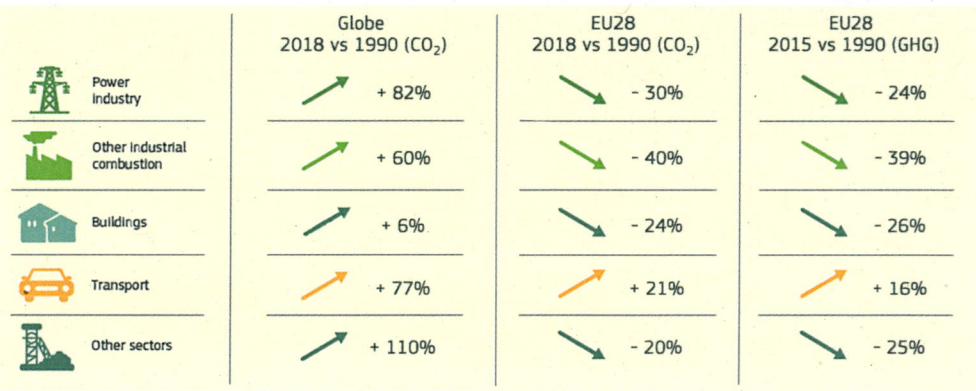

	Globe 2018 vs 1990 (CO$_2$)	EU28 2018 vs 1990 (CO$_2$)	EU28 2015 vs 1990 (GHG)
Power industry	+ 82%	− 30%	− 24%
Other industrial combustion	+ 60%	− 40%	− 39%
Buildings	+ 6%	− 24%	− 26%
Transport	+ 77%	+ 21%	+ 16%
Other sectors	+ 110%	− 20%	− 25%

Source: EDGAR website. tinyurl.com/y5eklj27. Accessed November 15, 2020. With permission of EDGAR.

would amount to an average rise in temperature of 1.1.°C[38] According to NASA, for example, global climate change has already had observable effects on the environment. Glaciers are shrinking, ice on rivers and lakes are breaking up earlier, plant and animal ranges are shifting, and trees are blossoming earlier. Effects that scientists in the past had predicted would result from global climate change are now occurring: loss of sea ice, accelerated sea level rise, as well as longer, more intense heat waves. Scientists are highly confident that global temperatures will continue to rise for decades to come, largely due to greenhouse gases produced by human activities from the past.

The Intergovernmental Panel on Climate Change (IPCC), which includes more than 1,300 scientists from many different countries, forecasts a temperature rise of 2.5 to 10°F (1.4 to above 5°C) over the next century. The IPCC predicts that increases in global mean temperature of less than 1.8 to 5.4°F (1 to 3°C) above 1990 levels will produce beneficial impacts in some regions and harmful ones in others. Source for both passages and further information: ⊿ climate.nasa.gov/effects

Some resources to deepen your knowledge of the effects of climate change effects and learn more through images, facts and numbers:

–"Global Climate Change: Vital Signs of the Planet" provides data, videos and images from the unique perspective of NASA.
⌦ climate.nasa.gov/   '

–"Our World in Data" with graphs, country data and further resources.
⌦ tinyurl.com/yy4p4w34

–The National Oceanic and Atmospheric Administration (NOAA) provides podcasts and videos on climate change and other topics, including videos and games for kids. It has a strong US perspective and certain simplicity. ⌦ tinyurl.com/y5r446gj

–"The Climate Change Guide" is a project by Laurent Cousineau, a young passionate Canadian professional. He explains, among others, the increases in natural disasters and addresses the topic of environmental refugees. ⌦ tinyurl.com/y47belvw

"The Sustainable Development Goals are a universal call to action to end poverty, protect the planet and improve the lives and prospects of everyone, everywhere. The 17 Goals were adopted by all UN Member States in 2015, as part of the 2030 Agenda for Sustainable Development which set out a 15-year plan to achieve the Goals."[39]

In addition to climate change, the 17 SGDs cover a wide variety of topics ranging from poverty and hunger to education, water and sanitation, sustainable cities, gender equality and decent work. This reflects an integral understanding of economic, environmental and social aspects.
The 17 SDGs are underpinned by a total of 169 targets and 232 indicators.

Sutainable Development Goals

The new 2030 Agenda for Sustainable Development aims to *"provide a shared blueprint for peace and prosperity for people and the planet, now and into the future ... The SDGs recognize that ending poverty and other deprivations must go hand-in-hand with strategies that improve health and education, reduce inequality, and spur economic growth—all with tackling climate change and working to preserve our oceans and forests."*[11]

The 2030 Agenda shows a close connection to the mindset of the Rio Earth Summit, yet the further development of broader priorities and the stronger engagement of different actors. Consequently, the SDGs today invite different actors to define their contribution; among those, companies are a major party. Companies increasingly apply the SDGs as a framework to define their strategic contribution to the global sustainable development agenda (see page 111).

BASIC CONCEPTS FOR SUSTAINABILITY

Let's take a look at some conceptual frameworks. From the many concepts, terms and definitions that exist, I would like to address five that have high business relevance or that are frequently used in the corporate world.

THREE PILLARS: WEAK OR STRONG?

Very often you will hear about the balance between economic, social and environmental concerns—the three pillars of sustainability, sometimes called "the magic triangle". The origin of this concept is not clearly attributable, as several actors had already adopted a similar integrative understanding in the early nineties. The German Association of the Chemical Industry (VCI), for example, mentioned in a discussion paper on political discourse in the mid-nineties, that economic, ecological and social aspects should be treated equally, as all three contribute to a sustainable future. The Global Reporting Initiative (GRI), today a major framework for communication on sustainability, specifically lists these three categories as topics and indicators for reporting (see page 158).

> **TIP**
>
> The 2020 SDG Report is easy to read and shows where our world stands with regard to the 17 SDGs. It includes the implications of Covid-19.
>
> ✈ tinyurl.com/y5rokp7c

When all three pillars are treated equally, this is sometimes referred to as "weak sustainability." In such a case, social, environmental and economic aspects have the same weight. For example, we would tolerate the depletion of natural capital such as forests because they would have been transformed into socially beneficial products or would have created jobs at the same time. This is our typical economic understanding and practice—we accept trade-offs. In the case of "strong sustainability," the understanding is that the boundaries are set more explicitly and treated as limited and largely non-replaceable by social or economic capital, particularly when it comes to natural capital. Natural capital—soils, forests, oceans, air, and so on—according to this understanding would form a limit or a corridor to our economic activity and social development.

In the academic world, it is argued that the three pillar concept does not provide sufficient direction as it lacks operationalization and prioritization. In practice, the three pillars have their limits, too: while they help organize topics at an initial level, they offer little support as to the specific management decisions to be made due to their shortcomings mentioned above. Therefore the concept is usually supplemented with other approaches such as materiality analysis (see pages 62ff.) and footprints (see page 58).

TRIPLE BOTTOM LINE & FINANCE

The bottom line typically calculates the financial result (profit/loss) resulting from a company's economic activity. Similarly, the triple bottom line aims to include a company's social and environmental performance alongside economic performance. This concept has a strong similarity to the three pillars and is often referred to in the business world, probably because the terms is familiar to business people and investors. The formula "people-planet-profit", with the underlying triple bottom line concept, was introduced in 1994 by John Elkington, a British consultant.

This formula came up against the same boundaries as the three pillar model with its lack of operationalization. It contributed, however, to an important development that would end up propelling the concept of sustainability within the financial sector. Investors would increasingly start to hold companies accountable for their performance in social and environmental spheres. The development of Environment, Social and Governance criteria, or so-called ESG criteria, in the global investment world, emerged and paved the way for non-financial reporting. You will sometimes hear the terms "ethical investment" or "sustainable investment." While these have a somewhat different focus, they all contribute to the growing awareness of sustainability matters in the investment world.

The topic emerged in the financial sector mainly because the economic impacts of social and environmental aspects were becoming more obvious and being awarded an economic value. Take, for example, the increase in environmental regulations (e.g., in the areas of wastewater treatment and chemical waste), resource scarcity, rising prices for natural resources, reputational risks arising from human rights violations, and the consequences of natural disasters. As economic relevance increased through rising costs and risks, ESG-related monitoring became more important in the investment world and, as a result, drove sustainability at respective companies and throughout whole industries.

An adopter of this trend is the international bank HSBC. As part of their sustainability approach *"Building a sustainable future: We help serve the needs of a changing world"*, HSBC focuses on three areas: sustainable finance, sustainable supply chains, employability and financial capability. Within sustainable finance, HSBC aspires to be a leader in financing, managing and shaping the transition to a low-carbon world. The bank has pledged to provide 100 billion US dolars of sustainable financing and investment by 2025, providing the money necessary to finance offshore wind plants and solutions for sustainable cities, among other things.[22] This commitment reflects a typical dynamic of linking business to specific aspects of sustainability: A bank uses its major asset—money—and decides to channel it into a future business—the low-carbon world—and in doing so addresses a major sustainability challenge—helping combat climate change and accelerate the momentum in other sectors such as wind energy.

TIP

A short analysis of HSBC shows how ESG became more important in the investment world.
tinyurl.com/y2r8fpoj

CSR—A DIFFUSE TERM

CSR is the acronym for Corporate Social Responsibility, which is sometimes called social responsibility or corporate responsibility. CSR is a frequently used term but with little consensus as to its meaning.

An academic study of 2011 listed 37 different definitions of CSR and concluded there was a need for a more common understanding.[42]

CSR is often used as a synonym for sustainability. The Walt Disney Company, for example, states: *"Our approach to corporate social responsibility is built upon the Company's long and enduring legacy of engagement in our workplaces and communities and our actions to protect the environment."* [43]

TIP

The EU green paper on CSR can be downloaded here.

⌁ **tinyurl.com/y6mo9vuu**

One source of the term CSR is a 2001 so-called green paper of the European Union titled "Promoting a European framework for Corporate Social Responsibility". According to this report, being socially responsible means not only fulfilling legal expectations but also going beyond compliance and investing in human capital, the environment and the relations with stakeholders. Companies should pursue social responsibility internally (e.g., maintaining employee health and safety, investing in workers' skills, adapting environmentally responsible production), as well as externally across their whole supply chain. The European Union and its Commission have broadened the focus even further, and nowadays companies are defined to be socially responsible by integrating social, environmental, ethical, consumer, and human rights concerns into their business strategy and operations. This understanding is aligned with the developments of the ISO (The International Standards Organization). The ISO has created an international standard for the social responsibility of private (corporate) and public sector organizations. The standard ISO 26000 establishes the following seven core subjects of social responsibility: Organizational Governance, Community Involvement and Development, Human Rights, Labor Practices, The Environment, Fair Operating Practices, and Consumer Issues. The guideline was released on November 1, 2010. Compared to other ISO standards, this standard serves as guidance and does not yet offer any type of certification.

There is another understanding of CSR as a synonym for corporate charity and other ways of giving back or contributing to society. The root of CSR's connection to charity lies in the understanding that companies have a social responsibility to share their wealth, donate their time and money to social causes, and support communities. We will take a deeper look at this area (see chapter 8). For CSR, it is important to define which understanding a particular person or organization has in order to find common ground.

FOOTPRINTS: MEASURE IT UP

Carbon footprint, water footprint, eco footprint, land footprint, fossil energy footprint vs renewable energy footprint, the footprint of bottled vs tap water ... Many different footprints can be measured, and we will refer to this concept several times. The methodology varies across the different footprints. Due to the complexity of the underlying data, footprints are often generalizing and approximate measures. Still, they are incredibly powerful for displaying the magnitude of impacts and detecting the key levers for improvement. We already mentioned per capita carbon footprints of countries (see page 50), and you can check out the per capita water footprints for different countries

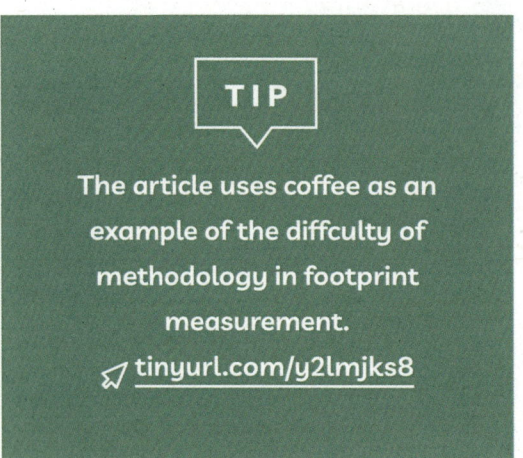

TIP

The article uses coffee as an example of the diffculty of methodology in footprint measurement.

✈ tinyurl.com/y2lmjks8

(see page 70). Let's look at different measures for the same categories.

A carbon footprint can also measure the amount of emissions produced by a certain activity. A roundtrip flight from Hamburg to Los Angeles, for example, measures up to 4.7 tons of CO_2, 2x of the whole carbon budget available for one person per year to live in accordance with the Paris Agreement. Water footprints can also measure how much water is needed across the whole value chain to produce a certain product. The standard approach to measuring the water footprint is to estimate the total amount of water consumed to produce, process, and transport products from their point-of-origin to the point-of-use and, based on the results, display how much water is embedded in a certain unit of the product.

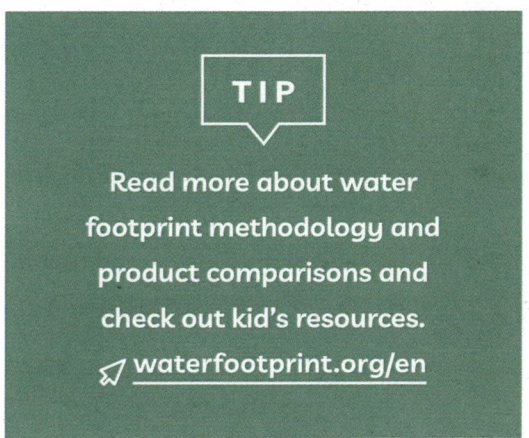

TIP

Read more about water footprint methodology and product comparisons and check out kid's resources.

⬈ waterfootprint.org/en

This displays, for example, that to make one pot of coffee (750 ml), it takes 840 liters of water across the whole value chain compared to only 90 liters for tea and 1,000 liters for a liter of milk.[44] Bottled water in comparison to tap water, for example, has a 5 times higher water footprint. This measurement approach also shows in a direct way, how manufactured products can add up to a larger water footprint. While a plain orange has a water footprint of 80 liters, a glass of manufactured, bottled orange juice adds up to 200 liters; the ratio is similar for an apple (125 liters) and a glass of processed juice (230 liters).[45]

Another perspective is to compare the footprint of various ways to reach the same goal. A typical example is carbon emissions in travel. A comparison shows the following: *"Overall, the most efficient ways to travel are via walking, bicycle, or train. Using a bike instead of a car for short trips would reduce your travel emissions by ~75%. Taking a train instead of a car for medium-length distances would cut your emissions by ~80%. Using a train instead of a domestic flight would reduce your emissions by ~84% ..."*.[46] It shows that footprints are usually a rough estimate. In reality, it depends on factors such as the length of the trip, the occupancy of the car, the mode of transport and the type of vehicle used. Still, these measures help to show the magnitudes and support conscious decision-making on the part of consumers.

Footprints are also an important tool applied at companies. In business, they are fundamental to finding out the main levers for reducing the carbon, water or soil footprint. Usually, companies calculate footprints through specific life cycle assessments (LCA). An example is displayed for a pair of Levi's 501 jeans. According to the analysis, one pair of jeans, for example, requires 3,781 liters of water, emits 33.4 kg of CO_2e and has a

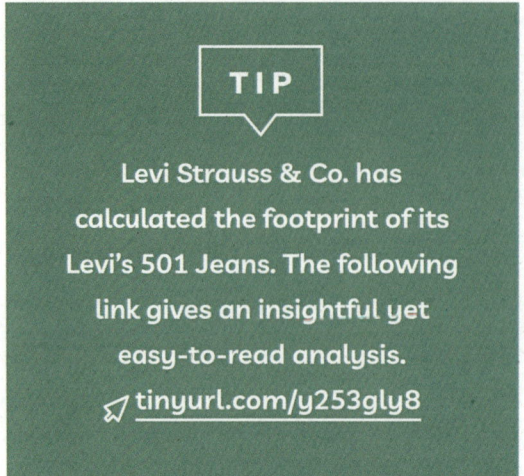

TIP

Levi Strauss & Co. has calculated the footprint of its Levi's 501 Jeans. The following link gives an insightful yet easy-to-read analysis.

tinyurl.com/y253gly8

land use of 12 square meters. Companies also try to identify key levers. In the specific case mentioned, the company also concluded that the water footprint significantly varies according to the washing habits of the consumer. Thus, a company could encourage customers to wash less or differently to reduce the water footprint.

Competitor Nike conducted a very detailed footprint measurement analysis across the whole value chain, including several upstream steps (so-called "tiers"). It showed that the entire steps of materials (growing, extraction, material manufacturing) are the most relevant from an environmental perspective.

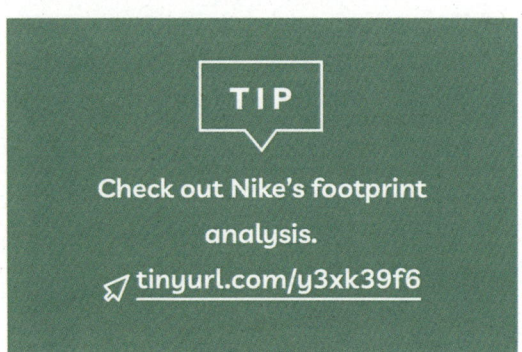

TIP

Check out Nike's footprint analysis.

tinyurl.com/y3xk39f6

Companies are increasingly called on to address the crucial areas in the value chain. In the dairy supply chain, for example, about 80% of the carbon footprint originates upstream in agriculture, compared to about 20% combined in manufacturing, retail and consumption, including packaging and recycling.

In automotive, about 80% of the carbon footprint occurs downstream during vehicle use and about 20% in sourcing and manufacturing.

WHEN TECHNOLOGY MAKES IT WORSE

In our world, where economies are built on growth and consumption, technology and innovation play a vital role in today's business. Both consumption and technology combine to form a little known but very important dynamic called rebound effects.

Technological developments luckily lead to more efficiency and often new market opportunities. Rebound effects occur when technical efficiency leads to increased demand in society, either directly for products or indirectly in other sectors of the economy. Although we become more efficient, we still end up consuming more overall, due to the fact that people's behavior is rooted in our paradigm of consumption and growth. The vicious circle of technology gets broken. We are all surrounded by rebound effects but are often unaware of what they are. The following are a few examples:

– The energy and water efficiency of electrical appliances have risen significantly, meaning we use less energy and water. However, households have a higher total number of appliances, such as washing machines, dishwashers, dryers, coolers etc., and use them more often.

– Heating costs in buildings have decreased due to better isolation and more efficient heating systems. But, while the average heating energy per square meter has declined, total heating energy has risen as a result of an increase in the average living space per capita.

– Passenger cars have become much more efficient over the past decade, possessing more power while using less gasoline. At the same time, the number of cars has grown as a household today typically has more than one car. Cars have also become bigger, and people are driving more with their cars on average.

– Technical devices such as laptops and smartphones have become dramatically more efficient, but we possess many more of them today. By 2020, there should be around 6.58 network-connected devices per person on average around the globe.[47]

Rebound effects are calculated to be between 10-50%, according to several studies.[48] Researchers and developers should (and increasingly do) not just look at the efficiency of certain technologies but also the behavior resulting from the availability of certain technologies. Researchers have observed that not only do direct rebound effects occur but also indirect rebound effects. Money saved on bills as a result of efficiency, for example, is spent on other consumption such as additional weekend travel or the next new device. Steve Sorrell, an energy policy expert at the University of Sussex, quantified this effect. He looked at three energy-saving actions: turning the heating down by 1°C, replacing car journeys under two miles with walking or cycling, and throwing away one-third less food. According to Sorrell, if you did these three things and then spent the money you saved in line with your typical spending patterns, the rebound effect would be 34%. In other words, 34% of the greenhouse gas reductions would be cancelled out as a result of the goods and services the extra money would be spent on.[49]

2.03 ——————— TRANSLATE IT INTO BUSINESS

Sustainability has been described so far as a vision for living in this world together. Such a generic principle has to be made practical. There is no such thing as "the sustainability". It must be defined within the specific context of a company, industry or product. A practical approach to defining sustainability for a company or industry is a materiality analysis. We will also look, however, at what sustainability means for three different products and services, respectively.

MATERIALITY: WHAT IS RELEVANT?

Starting with the three pillar or triple bottom line concept that breaks down sustainability into three areas (economic, social, environmental), a specific business context needs to specify which topics have the most impact and relevance. A materiality analysis helps to do this. It collects, evaluates and prioritizes issues for a company. To identify the issues of a company, the first aspects to look at are the industry it operates in, the specifics of the value chain and the business model. In addition, the topics raised by critical stakeholders such as environmental and civil rights organizations, the requirements of customers and suppliers, the activities of competitors, and the regulatory developments and future megatrends are all analyzed. Much of the information can be collected through extensive research on sustainability issues for the specific company setting. Often, the analysis is underpinned by stakeholder surveys, especially for prioritizing to the most relevant issues.

After conducting a large number of materiality analyses for companies, I know that the initial number of environmental, social and economic issues in large corporations can easily sum up to roughly a hundred topics. These have to be evaluated, structured, bundled and prioritized. They will eventually be reduced to about 5 to 20 material topics, depending on a variety of considerations, and later translated into strategic tasks. A materiality analysis is not only helpful for the companies but has become a prerequisite today for different management and reporting systems.

Take Microsoft, for example. In a materiality assessment, it defined the following ten topics as the most relevant for their business. Note that these topics very specifically reflect the business's characteristics: Accessibility/Applying technology for environmental and social good/Climate change and energy/Closing the broadband gap/Ethical business practices/ Human capital development/Human rights/Privacy and cybersecurity/Responsible AI (Artificial Intelligence)/Skills and employability.[50]

Often, the topics are depicted in a "materiality matrix" that prioritizes issues. It visualizes a company's most important sustainability matters in terms of their relevance and impact. To see how different material topics can be at different companies, here is an example of two companies who, within their own very different industries, are among the world's largest. Compare their issues as well to those of Microsoft.

Unilever is a consumer goods company with brands in food and beverages (e.g., Lipton, Magnum, Knorr), home care (e.g., Domestos, Coral) and beauty and personal care (e.g., Axe, Dove). LafargeHolcim is the world's biggest producer of building materials such as cement, concrete and infrastructure solutions. The different material issues reflect the different business impacts.

Take the materiality matrix of LafargeHolcim. Health & safety is highly relevant, as 75,000 people around the world work in factories and are likely exposed to accidents. Their protection is a priority for the company. Accordingly, local communities play a strong role, and impact is created, for example, through building houses in underserved communities.

LafargeHolcim Materiality Matrix

The issues that we will focus on in the next 3–5 years in order to create value for all stakeholders.

KEY

Focus

Monitor and manage

Maintain

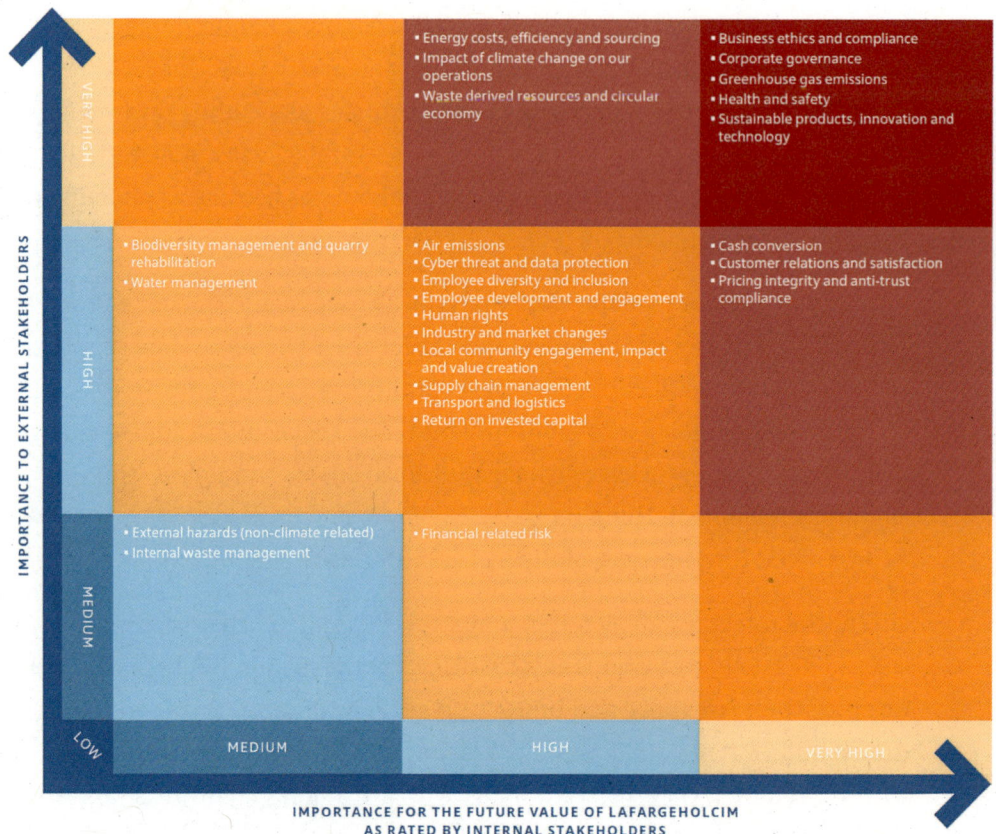

Source: LafargeHolcim Integrated Annual Report 2019. With permission of LafargeHolcim.

Greenhouse gas emissions and energy management are two of the most relevant topics. Due to its energy-intense production processes for building materials, this industry is a major emitter. Recently, Lafarge-Holcim made a net zero climate pledge. The importance of this is reflected in the opinion of one of the stakeholders: "We are delighted that Lafarge-Holcim has joined the group of over 290 industry leaders committed to a 1.5°C future. As the largest player in one of the most carbon-intensive industries, LafargeHolcim's leadership demonstrates that a net zero economy is within reach."[51] The full pledge can be read here: ✎ tinyurl.com/y2qz4vw7

In the matrix, you can also see the importance the company assigns to certain topics. For example, focus topics are a higher priority than monitored topics. A materiality matrix is typically updated every two to five years to include newer developments within the company or topics raised by stakeholders.

Unilever Materiality Matrix 2019/2020

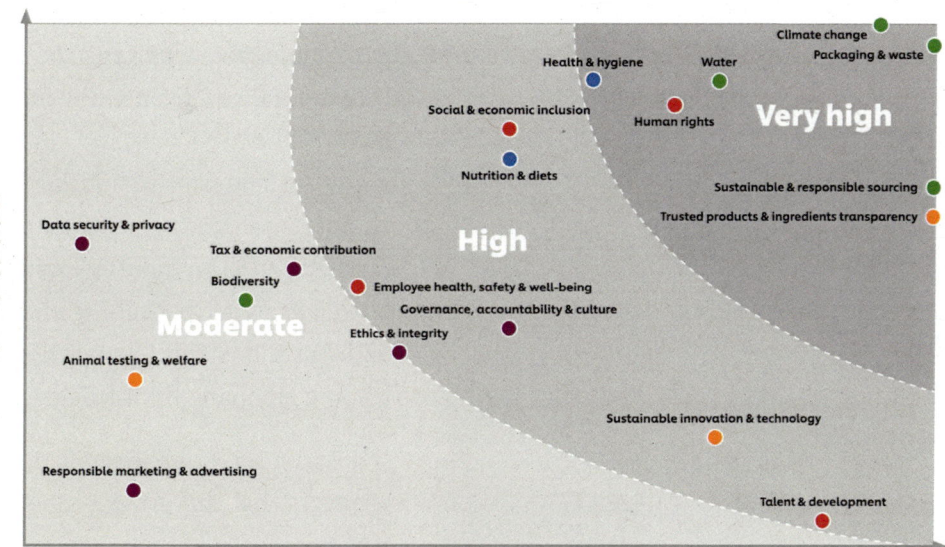

Source: Unilever website. tinyurl.com/y6lyvucj. Accessed November 28, 2020. With permission of Unilever.

The business impact category Packaging & waste, for example, is a major topic for a consumer goods company such as Unilever. Sustainable & responsible sourcing applies especially to the agricultural products in Unilever's food segment. Health & hygiene concerns primarily to its non-food business.

To summarize: Across industries and companies, different sustainability topics are relevant. They must be analyzed and pinned down. Ideally, the materiality analysis forms the basis of the strategy (see pages 111ff.).

HOW ABOUT AN INDUSTRIAL FOAM?

Let's switch from the company level to the level of products and services, starting with a product you are very likely surrounded by without knowing it. "Armacell is everywhere" is the slogan of the company that invented flexible foams for equipment insulation and is a leading provider of engineered foams. Simply stated, Armacell contributes to solutions that conserve energy, reduce noise levels, keep food fresh or add comfort and safety. Insulation materials are used, for example, in a wide range of commercial and residential equipment for heating, acoustic and vibration solutions, in industrial equipment manufacturing for pharmaceuticals, power generation, as well as in the packaging and logistics industries. Armacell is a leading company in this field.

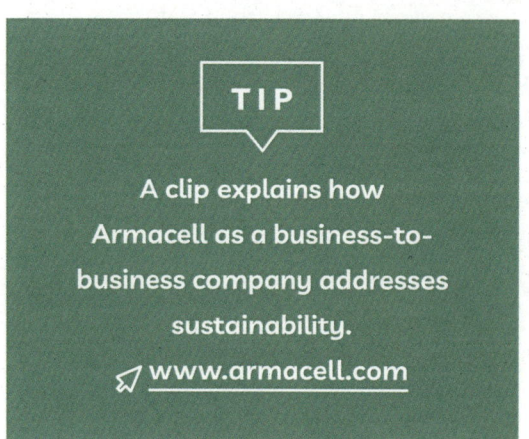

TIP

A clip explains how Armacell as a business-to-business company addresses sustainability.
www.armacell.com

Industrial foams are chemical products that can provide a sustainability value-add in its application. For example, Armacell has developed component foams that facilitate lightweighting in the automotive industry, which is forecasted to lower this industry's CO_2 emissions by up to 40% by 2050. As a PET pioneer investing a decade of research in developing recycled PET foam technology (rPET), Armacell has a further application that closes resource cycles. In 2015, it invented the first and, as of now, only 100% recyclable thermoplastic foil product for recycled PET. The large-scale production of 3D-shaped PET foam cores followed in 2018. Foam core materials using rPET technology generate over a third less CO_2 than those made of virgin PET, while PVC foams, a competing technology, typically cause twice the volume of CO_2 emissions. Until today, more than 1 billion PET bottles have been re-used by Armacell to manufacture PET foams that are applied in the construction, transportation and wind industries. The example shows that sustainability often comes in a very specific, complex way and involves advanced technological approaches.[52]

A SUSTAINABLE CUP OF COFFEE

Coffee is a great example of how many different attributes can apply to a single product—each of them being a facet of sustainability, but none of them "encompassing them all".

Growing coffee is water-intensive. As the water footprint comparison to tea revealed, growing methods using less water are preferable. One such method is **shade-grown coffee**, which additionally has a significant positive impact on biodiversity and is therefore sometimes sold as **bird-friendly coffee**. However, due to current standard industrial production practices, shade-grown coffee is a niche method in most regions and applied much less than organic production.

Organically grown coffee prohibits the use of pesticides and chemical fertilizers, which are heavily used in conventional growing. Thus, the organic method fosters the health of soils, ecosystems and the workers who don't have to apply them. **Zero-emission roaster** and **emission-free coffee** focus on offsetting the carbon dioxide associated with the growth, production and logistics of the coffee but don't necessarily include any of the attributes mentioned above. **Fairtrade** is an attribute underpinned by a certification–either on its own or often combined with organic coffee. The focus of Fairtrade is on promoting justice as well as on reducing poverty, as it guarantees a surplus income for the coffee growers.

The certification **Rainforest Alliance** provides basic social and environmental standards in the growing countries and is usually applied to the mass market. Compared to other sustainability standards, Rainforest Alliance is more like a baseline or minimum standard. Roasters and smaller coffee shops in cities around the world have established **special partnerships with cooperatives** in grower countries. This can include long-term contracts, higher prices, projects supporting the cooperatives and their communities, the investment of money and sometimes the development of skills by building schools and hospitals or creating job opportunities. Most of these are individual partnerships but not certified or independently accredited.

A further attribute to be considered includes **sustainable packaging** such as the use of less plastic and more natural materials such as paper. It should be mentioned that at to-go coffee shops, the disposable coffee cups add a significant drawback and an environmental challenge to your cup of coffee, even if it is Fairtrade or organic.

The different attributes mentioned demonstrate that there is no such thing as THE sustainable coffee, and features do not easily compare. This also implies that individual preferences matter: some customers care more about environmental topics, others about social topics. Within these preferences, different aspects are highlighted through existing approaches.

> ### TIP
>
> Avoid to-go cups. At least 16 billion disposable coffee cups are used each year. In Germany, it's 3 billion per year—320,000 cups per hour. [53] In the UK, it's 2.5 billion, of which less than 1% is recycled.[54] As disposable coffee cups are coated with plastic and contaminated with drinks, they usually cannot be recycled by standard recycling plants.

WHAT MAKES BANKING SUSTAINABLE?

Compared to coffee, it is more difficult to match clear, established attributes to banking. A starting point is ethics, which became very important following the financial crisis and several scandals and dubious transactions involving major banks. Ethics defines a baseline. Each person would probably want its bank–for private or business purposes–to discourage fraud and not engage in shady business operations. Additionally, most would not want their bank to invest in business practices that harm the environment (for example, projects that destroy tropical forests) or people (for example, those financing the armaments industry or speculating in high-risk investments). This sounds reasonable, yet it is challenging to implement and monitor. A complex structure of institutional investors, ratings and ranking agencies, compliance boards and supervisory authorities has evolved over the past decade to raise the bar in the area of ethics. Consequently, a sustainable banking service would be one from a bank with ethical business practices in place who was also transparent about its business and how it handles ethical issues.

The main question associated with sustainability is what the bank does with the money it manages. Here again, individual consumer preferences play a role. HSBC, as we saw, channels a significant amount of their money into customer segments that foster solutions combating climate change. The cooperative GLS Bank in Germany invests in social infrastructure by financing kindergarten projects and environmental assets such as renewable energy. Other banks invest primarily in the local economy and support small businesses. The money managed by banks exerts the highest societal impact–for the good or not so good. The environmental footprint of daily banking operations is less significant compared to other industries. Still, any effort to reduce energy, paper and other resources counts. Providing employees with a good work environment is another aspect that reflects an attitude of responsibility that translates into trustworthiness and a solid reputation. For a bank, these are decisive. Rather than talk about a sustainable banking service, we would consider our bank to be working ethically and sustainably. The industry has been gradually starting to embark on more sustainable pathways over the past decade.

Let's sum up at this point. Throughout the book, you will find many applications for translating sustainability for a business or product. Remember, specifying sustainability and the main levers in an industry or company is challenging yet necessary. What it means for a specific business must be detected from the business model, the value chain the corporation is operating in, and the overall industry dynamics.

There is no textbook definition for what sustainability means for a specific product or service; instead, it has to be determined for each individually based on concrete attributes. Consider, for example, what suitable sustainability attributes would be for a car, toilet paper, an apple, cereal, a dishwasher, a mobile phone, a streaming service, hair coloring at the hairdresser, a family's summer vacation? What is it for even more complex items such as an airplane, a chemical substance in an industrial process, a building in a megacity or an intelligent data system? The attributes of sustainability differ widely and are not always easy to detect.

APPLICATION

Take a quiz!

1. ———— True or false? Sustainability …
a) … has clear criteria for each product/
industry, area of life.
b) … is a general principle of how
we aspire to live and do business.
c) … seeks a balance between economic,
social and ecologic parameters.
d) … means climate change.
e) … is taken care of by politicians.

2. ———— How much water is consumed? Match
the countries with their numbers. Water
footprint per capita per day (in liters)[55]

United States	7,800
Brazil	5,600
Australia	6,300
Indonesia	3,100
China	2,900
Germany	3,900
Mozambique	3,100
Laos	2,900
Congo (DRC)	1,500

3. ——— To each product, match two attributes that would qualify for improved sustainability of the respective product.

Product	Attributes	
Cell phone	Vegan	Conflict-free minerals
	Locally produced	Regional
Orange juice	No meat	Low resource use
Liquid soap	Low-emission	High-quality
	Uses recycled paper	Fairtrade
T-shirt	Reusable	No microplastic
Car	High-effciency	Homemade
	Natural	No additives
Event	Long-living	Plastic-free
Printer	Palm-oil free	Low energy use
	Organic	Carbon offset
Pesto	Accessible	Uses renewable materials

UNDER-STANDING DRIVERS

Back in the early 2000's, I started to consult on sustainability and had several bizarre conversations with corporate executives along the lines of "Sustainability isn't relevant." "Customers don't care." "I don't understand it (so, it doesn't matter)." "We have more important things to do." At that time, some people were reluctant and even arrogant towards this passionate young woman, with looks like "What could you, as a newcomer, be able to tell me, a big executive, about my business and the gloom that lies ahead?"

Today, I am thankful for these encounters and experiences. They have taught me to argue, analyze barriers, and find solutions. The tipping point for sustainablity was reached some years later, not really like a big bang but more like impetus coming from all sides. Now, looking ahead, I believe we need even more strategic foresight based on passion and deep insight.

─────── # RECOGNIZING STAKEHOLDERS

A key to understanding what drives sustainability is the concept of stake-holders. The word "stakeholder" is analogous to the term shareholder, which is a party with ownership in a company through the possession of shares. Instead of shares, stakeholders have "stakes" in organizations, which is another way to say claims, interests, concerns or demands.

A BRIEF LOOK AT STAKEHOLDERS

A simple description: Stakeholders (also: stockholders) are those who may be affected by what a company does, or they have an impact on the company and can therefore exert influence on its actions. Neighbors of a production plant, for example, are external stakeholders who could be negatively affected by traffic, noise, or other emissions. Employees of that same plant are internal stakeholders who are positively affected when the company pays fair wages and takes care of the employees' health. Regulatory bodies who approve a desired expansion of the production plant could ask the company to fulfill certain conditions in return—for example, to minimize emissions to a certain level or employ people with disabilities. An environmental organization that wants to protect the forest around the plant could provoke delays in the approval or even prevent the expansion altogether. As a critical, external stakeholder, they would force their interests upon the company, which could negatively impact it.

The concept of stakeholders is, meanwhile, a solid part of organizational management and, increasingly, an integral part of sustainability management at companies. With his book "Strategic Management: A Stakeholder Approach", published in 1984, US philosopher and professor of business administration R. Edward Freeman is considered the "father of stakeholder theory". Stakeholder theory addresses morals and values in managing an organization and, among other things, includes perspectives about the roles and responsibilities of different entities within a society.

Stakeholders can be persons, groups of persons or organizations. This accepted definition reveals an anthropocentric perspective that does not give plants, animals, geology and the alike a voice as stakeholders, but only an instrumental value in relation to human groups or individuals. Many non-governmental organizations, as well as policymakers and other groups or individuals are advocates of the environment and social causes. In reality, some issues may remain excluded if there is not enough public awareness or strong advocacy in their favor. This is one reason why activists and other actors who critically observe development and stand up for the rights of other beings are so important and valuable.

Employees, suppliers, local communities, creditors, media, policymakers, non-governmental organizations and research institutions are among the common categories of stakeholders. Freeman also included competitors–which, as can be seen by the published stakeholder lists, is less common in practice but highly relevant for a corporation when analyzing industry dynamics and benchmarking in sustainability.

WHO IS RELEVANT?

A common approach is to define the specific stakeholders of a company, identify their concerns and their impact in order to examine how to interact with them through stakeholder management. The following table lists the stakeholder groups of four companies as they are described in their disclosure on websites or in sustainability reports. It can be noted that some stakeholders repeatedly appear and that different segmentations are applied that reflect the variety of the companies' businesses and their approaches.

For listing stakeholder groups in a sustainability report or conducting a materiality analysis, such a broad segmentation of stakeholders is usually sufficient. As I learned in several projects, the specific breakdown of stakeholders depends on the purpose. For example, for a specific risk analysis of a particular sustainability topic, a company usually needs profound segmentation of the relevant actors and their specific concerns, as well as their likely power and impact on the company.

COMPANY	NESTLÉ [56]	MICROSOFT [57]
STAKE-HOLDER GROUPS	- Academia - Communities - Consumers and the general public - Customers - Employees and their representatives - Governments - Industry and trade associations - Intergovernmental organizations - NGOs - Reporting agencies - Shareholders and the financial community - Suppliers (including farmers and smallholders)	- Customers - Investors - Employees - Suppliers - Civil society/non-governmental organizations - Communities - Industry coalitions and public-private partnerships - Policymakers - International Government Organizations

COMPANY	DAIMLER [58]	BAYER [59]
STAKE-HOLDER GROUPS	**Primary** - Shareholders - Employees - Customers - Suppliers **Communicating regularly with** - Civil groups such as NGOs - Associations - Trade unions - Media - Analysts - Municipalities - Residents in the communities where we operate - Representatives of science and government	**Partners** - Suppliers - Customers - Employees - Associations - Universities/schools **Financial market participants** - Investors - Banks - Rating agencies **Social interest groups** - General public - NGOs - Local communities - Competitors **Regulators** - Lawmakers - Politicians - Authorities

Own presentation

WHO HAS AN IMPACT?

In the 2019 study of the UN Global Compact together with Accenture Strategy, top executives of leading companies across different industries were asked about stakeholder impact, among other things. A specific question was: *"Over the next 5 years, which stakeholder groups do you believe will have the greatest impact on the way you manage sustainability?"* It draws interesting insights to find out which stakeholders are considered most in which industries. The following percentages reflect the answer:[24]

– Consumers rank highest in personal and household goods (79%), food (70%), telecommunication (70%), retail (67%), chemicals (58%) and media (57%), and slightly over 50% in the other analyzed industries such as automobiles and parts, industrial goods and services and real estate.

– Employees are seen as the most important by some industries such as technology, financial services and construction materials. In real estate and media, employees have the same importance as consumers.

– Banks and insurances see regulators as their most significant stakeholders. Banks, additionally rank boards equally high. The basic resource industry considers the opinion of governments most, as does the travel and leisure industry. Oil and gas has a majority of 71% for communities.

3.02 ——————— TRIGGER EVENTS

We all know what impact certain events can have—just think of 9/11. In connection with sustainability, several incidents have occurred in the past that have triggered change in companies and industries. Let's look at three different events from diverse industries. All of them have a variety of difficult questions that still need to be answered.

A GREAT DISASTER

How does Bhopal sound to you? At an Indian plant producing pesticides, one of the largest industrial disasters on record happened in 1984 due to a leak. Estimates of the death toll ranged from 3,700 to 16,000. Over 500,000 people were exposed to the highly toxic substance that made its way into the area surrounding the plant. It was operated by Union Carbide India, an Indian company that shared stock ownership with, among others, the Union Carbide Corporation, which is today part of US-based Dow Chemical.

TIP

More about Responsible Care®.

⋈ tinyurl.com/y43y76y2

Websites examples for countries, such as the US.

⋈ tinyurl.com/yymtascz

As a consequence of Bhopal, the chemical industry worked to develop and globally implement Responsible Care®. This voluntary commitment of the global chemical industry seeks to prevent such an event from happening in the future. Responsible Care® includes improving process safety standards, community awareness and emergency preparedness. In its Guiding Principles, companies commit to the improvement of environmental, health, safety and security (EHS&S) performance for facilities, processes and products throughout the entire operating system. They also commit to open and transparent reporting and undergo mandatory headquarters and facility audits to certify their performance. Started in 1985, it now runs in 67 countries and includes 96% of the largest chemical producers in the world. Workers and communities are protected by enabling governmental bodies to implement industry best practices through regulations. Responsible care® is today like a "sustainability housekeeping" in the chemical industry.

Though the plant was closed in 1986, Bhopal caused environmental and health problems that are still present today. Contamination, for example, continues to be a huge problem in the region. Complex issues such as an

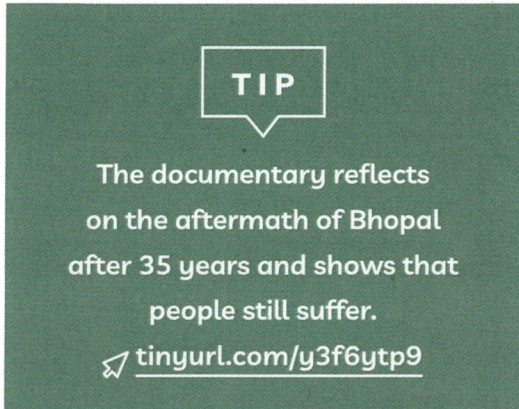

TIP

The documentary reflects on the aftermath of Bhopal after 35 years and shows that people still suffer.

tinyurl.com/y3f6ytp9

industrial disasters cannot and shall not be easily evaluated. A legal agreement of Union Carbide with the Indian government was settled. Dow Chemical has distanced itself from responsibility as it never owned the plant or was involved in the incident, as it describes on its website.

It is vital to raise the question of a company's future legacy when doing business, especially in vulnerable places. To which extent does it have responsibility for the people, the environment and the long-term effects? How does a company practice responsibility when people are not backed up by their political systems? Is a successor like Dow Chemical still responsible? To what extent? This is one of the very difficult cases.

A POWER STRUGGLE OFFSHORE

In 1995, an incident became an issue of public concern, especially in the UK and Germany. Brent Spar, a North Sea oil storage and tanker loading buoy operated by Shell UK was planned to be disposed of in the Atlantic. Environmental activist organization Greenpeace organized a worldwide, high-profile media campaign against this plan, occupying Brent Spar for more than three weeks with some 25 activists, photographers and journalists.

Shell had carried out an environmental impact assessment in advance in full accordance with existing legislation and argued that sinking the platform was the safest option, both from an environmental and an industrial health and safety perspective. Confronted with public criticism and politicians calling for boycotts, for example, across Shell service stations in Germany, Shell eventually abandoned its plans to sink the platform. The Brent Spar operation cost Shell several million pounds and tremendous damage to its reputation.

TIP

The history of the Brent Spar dynamic can be traced by this extensive article.

tinyurl.com/y6kwjmsp

Why? Shell underestimated public pressure and the long-term interests of powerful stakeholders. Environmental activists wanted to avoid having future disposals of oil facilities offshore, with Brent Spar as a precedent. They wanted to establish onshore disposal as a common practice in the oil and gas industry. As a powerful stakeholder, Greenpeace, who was later also criticized for some unfair measures in addition to the illegal occupation, got other powerful stakeholders (politicians, media, consumers) aligned to oppose to Shell's plans. Let's remember, although Shell's practice was legally correct and approved–the social license to operate in this specific case was temporarily withdrawn (see page 22).

In late 2019, Greenpeace activists again protested against Shell's offshore infrastructure with the demand, *"Clean up your mess, Shell"*. Once more, the complete removal of the oil platform infrastructure is at stake as significant offshore waste from oil operations is still out there in the seas.[60]

Here too, complex questions remain. How does the analysis of the social and environmental impacts of Shell compare to the demands of environmental advocates? Who decides on the advocacy for the oceans? What is a company's responsibility when legal conditions allow certain practices? How can responsibility for disposal be better enforced?

ANOTHER TRAGIC ACCIDENT

In 2013, the Rana Plaza collapse in Bangladesh caused the loss of more than 1,100 lives. It was the deadliest disaster in the history of the clothing manufacturing industry. The eight-story commercial building in the Dhaka District included several major garment factories manufacturing for brands such as Walmart, Gap, Benetton, and Primark. The garment industry is Bangladesh's biggest industry yielding 28 billion US

TIP

A video summary of the change in Bangladesh's textile factories in 2018, 5 years after the crash.
tinyurl.com/y65txs2z

An article highlighting the situation in 2020, at the 7-year anniversary of the Rana Plaza collapse. Due to the Corona crisis, fashion brands have mass canceled orders already produced or in production with severe impacts on workers.
tinyurl.com/y4qn9etf

dollars per year. The minimum monthly wage for garment workers in Bangladesh is 82 US dollars.[61]

The Rana Plaza building was constructed with substandard materials on swampy ground. The collapse of the complex once more put pressure on companies buying clothing from Bangladesh to provide for greater safety in factories. As a consequence, new unified workplace safety standards for clothing factories were installed and included inspections, training and regulation for spacing, escape and access, among others. The higher standards would apply to factories producing for brands from Europe and the US. They maintained their contracts with Bangladesh's factories and paid a share of the upgrade and maintenance costs for a two-year period while overseeing the inspection. Other factories, especially subcontractors, would not be covered by the new standards.

Who is responsible for the tragic event and the improvements? Is it the fashion retail industry or other actors in the supply chain? Is the best approach to merely ensure that one's own factories operate according to higher standards? Is it a responsible business practice to cancel orders "across-the-board" in times of crisis when there are vulnerable people involved? We will look at the industry in more detail and will see that, here too, there are no easy answers. Change requires numerous stakeholders acting together to improve conditions for workers. When it comes to this topic, we are all in.

KEY DYNAMICS FOR CHANGE

One of the key convictions you will repeatedly find throughout this book stems from my experience in consulting and academia that sustainability is very particular and specific to the context. Thus, dynamics in industries differ. In the case of trigger events, dynamics resulted in reactions to major incidents. Let's look at three further industries—the retail industry, the energy sector and the automotive industry—to further understand how dynamics likely emerge and accelerate.

THE WAKEUP OF THE RETAIL MARKET

In 2007, I was asked to write an article on "Strategic CSR in the German retail market" that was published in a compilation. I had worked on some client projects with retailers and had done academic research on sustainable innovation in the food industry for my MBA in Sustainability Management. From the analysis of the food industry, it became evident that a major barrier to more sustainable products reaching consumers were the retailers: the German retail market, consisting of a few powerful players, functioned as a type of "gatekeeper for sustainability".

It was a time when retailers cared about many things, prices, quality, branding—but not about sustainability. Not a single company had a program nor a sustainability officer. They barely addressed environmental or social topics in their own operations. Why then would they care about more sustainable products on their shelves? In those days, the frontend to the customer was of little relevance. Repeatedly, almost mantra-like, managers said, *"Customers don't care, they are not willing to pay more."* No other stakeholders were putting pressure on the retailers or the system, as we saw in other industries such as the oil and gas and chemical industries. As a consequence, sustainability was not on the agenda of the retail industry, or of the industries connected to it (e.g., food suppliers, other consumer good suppliers, packaging companies, logistics companies).

Ten years later, the landscape looked completely different. And, in 2020, it appears as if retailers have never cared about anything more than sustainability. A ban on plastic bags, more sustainable packaging, in-house organic brands, vegan product ranges, grass-fed dairy products, regional products and projects, measures to reduce food waste, sustainable sourcing, labels about animal welfare, transparency about origin, pledges to renewable energy, programs to improve the working conditions of supermarket staff, initiatives on biodiversity and bee friendliness. All retailers are active in sustainability and have used their significant market power. Over the years they have put increased pressure on their value chain, including restrictions on antibiotic use in cows; requiring GMO-free feeding of cattle in the dairy sector in Germany; sourcing raw materials such as coffee, cocoa and palm oil according to international standards; and applying new regulations to packaging.

Once retailers arrived at the point where sustainability mattered, they sped up by driving it forward in their own activities and especially in the value chain. The tipping point was reached through a variety of societal and regulatory drivers. Still, it is an industry with a strong herd mentality—so, a major push was given in the late 2000's when, in some lead markets, usually a forerunner started to sincerely commit to sustainability.

M&S in the UK for instance published "Plan A", a framework including 100 points to improve sustainability. It's called Plan A, they argued, because, with just one earth, there is no Plan B. In the US, Walmart launched a strategy to reduce the company's impact on the environment through three ambitious goals: to be supplied 100% by renewable energy, to create zero waste, and to sell products that sustain our resources and the environment. Walmart's CEO announced these goals 2005 in a speech broadcasted to all back then 1.6 million employees in all 6,000 stores and shared it with some 60,000 suppliers worldwide.[62]

In Germany, retailer Edeka, who had already been engaged in sustainability-related activities for several years, mainly through its partnership with the environmental organization WWF, had put together a sustainable product portfolio including fish and fruits. But the follower effect started when Rewe initiated some activities triggered by an ambitious CEO with an affinity for sustainability. Rewe decided to focus on some low-hanging fruit and be less bold than M&S or Walmart. But this move eventually triggered a chain reaction in the German retail market. Not only did the retail industry change but also the value chain, especially in the case of food producers and other consumer goods brand companies.

TRANSFORMATION OF THE ENERGY SECTOR

Societies consume large amounts of fuel, making the energy sector a crucial part of economies. It includes all industries involved in the production and sale of energy, including fuel extraction, manufacturing, refining, and distribution. As it is still dominated by the fossil fuel industries, the transformation towards a greener energy sector has accelerated. The two major drivers—policymaking and transparency—are highlighted below.

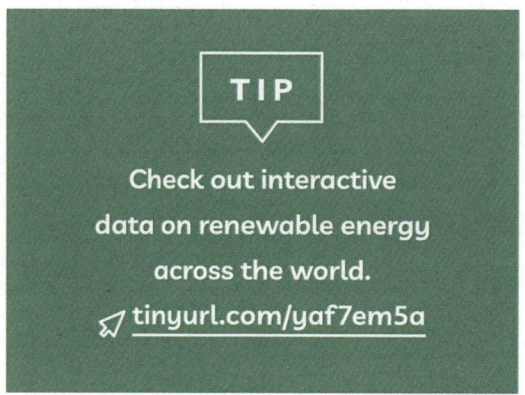

TIP

Check out interactive data on renewable energy across the world.
tinyurl.com/yaf7em5a

Worldwide, the production of renewable energy almost doubled between 2009 and 2018, according to data from the International Renewable Energy Agency. The share within the global power generation mix has risen steadily since the end of the 2000's and was almost 26% in 2018. Most of this growth comes from new wind and solar capacity, but the highest total share comes from hydropower. Renewables will continue to penetrate the global energy mix, with solar and wind generation expected to increase by a factor of 60 and 13, respectively from 2015 to 2050, as shown by a McKinsey analysis.[63]

According to Enerdata, this share in 2019 was 41.2% in Germany. Some countries are leading the crowd with shares of more than 50% of renewables in total energy production.

Share of renewables in total energy production (%), 2019[64]

Norway	Brazil	New Zealand	Venezuela
97.6	82.3	81.9	73.9
Colombia	**Canada**	**Sweden**	**Portugal**
72.6	64.9	58.7	55

Own presentation

A dramatic fall in solar and wind development costs over recent years due to technological advances contributed to the expansion of renewables. But experts agree: The growth of renewable energies was significantly supported by policymaking. It has gained momentum in the past decade in various countries helped by a supportive policy and regulatory framework. Government support in terms of incentives and subsidies played a crucial role in sustaining renewable development.[65] The most significant barriers to the widespread implementation of large-scale renewable energy and low-carbon energy strategies are primarily political and not technological.

TIP

Analysis and studies on the future of the energy sector and implications for different industries are available on the website of McKinsey.
⬀ tinyurl.com/y6ock3qm

Europe forged ahead in the 2000's. Passing energy policies that included prioritizing renewables and setting goals, creating financial incentives and shaping industrial policy, in addition to supporting the buildup of renewable energy clusters, fostering innovation and grassroots initiatives like citizen's solar parks were all among the political measures.

A mega project "Desertec" was initiated by different stakeholders with the goal of generating solar power in the Saharan desert to provide green energy to Europe. The project did not take-off adequately and, according to experts, the reason was the lack of

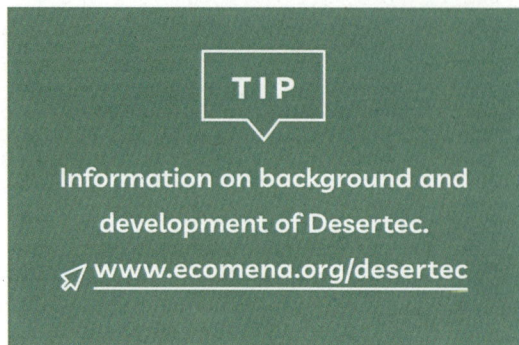

TIP

Information on background and development of Desertec.

✈ www.ecomena.org/desertec

ownership in northern Africa. With no efforts to foster sustainable development in the region, it was not possible for a megaproject like Desertec to flourish. There has been a fresh start to the project recently, with an adapted focus and stakeholder set. A vivid outcome is the biggest solar power plant in the world operating in the Sahara desert of Morocco. At 3,500 soccer fields in size, it's about the size of San Francisco.[66]

The 2015 Paris Agreement on climate change provided another boost and motivated several more countries to develop or improve their renewable energy policies and set targets—which now exist for more than 170 countries. Today, China is leading the way in renewable energy, being the world's largest producer, exporter and installer of solar panels, wind turbines, batteries and electric vehicles.[67]

TIP

The CDP is the leading platform to display progress on reducing the impact on carbon, water and forests, with good public data and reports.

✈ www.cdp.net/en

A second driver to the sector was the parallel push for transparency. In the energy sector, aside from ESG for listed companies, transparency was mainly propelled forward by the Carbon Disclosure Project (CDP). Launched in 2000, the CDP has developed into a reputable actor over a period of just a few years. The CDP's scoring drives corporate transparency and helps guide, incentivize and assess environmental action throughout industries, especially those that are carbon-intensive. Data on energy consumption, carbon emissions and goals for climate protection, among others, are collected by the CDP on an annual basis. This data is voluntarily provided by companies. However, because institutional investors use this data for evaluating ESG performance, pressure for disclosure is being put on the corporate world. In 2019, over 8,400 companies disclosed through CDP—a significant 20% increase over the previous year. Reporting companies now represent over 50% of global market capitalization. Because companies in the energy sector are a significant source of carbon emissions, yearly reporting enforces disclosure, creates transparency, allows for comparability and raises the bar for climate ambition. A total of 250 companies from the oil and gas industry and over 270 from the electric utility industry currently disclose through the CDP.[68]

CAN CARS BE SUSTAINABLE?

The automotive industry provides millions of jobs and offers mobility to people around the world. Transport systems have significant impacts on the environment, accounting for 20 and 25% of world energy consumption and carbon emissions and are still rising fast. Road transport is a major contributor to local air pollution and smog and, thus, causes social problems. The United Nations Environment Programme (UNEP) estimates that each year, 2.4 million premature deaths from outdoor air pollution could be avoided. *"The social costs of transport include road crashes, air pollution, physical inactivity, time taken away from the family while commuting and vulnerability to fuel price increases. Many of these negative impacts fall disproportionately on those social groups who are also least likely to own and drive cars. Traffic congestion imposes economic costs by wasting people's time and by slowing the delivery of goods and services."*[69]

> **TIP**
>
> A case study reports 20 years of sustainability progress in the UK automotive industry, showing the balance of also economic aspects.
> tinyurl.com/y5vodpnv

The automotive industry, with its focus on technical engineering and new designs, is very innovative. Over time, sustainability increasingly became involved. Early on this was triggered by regulation, for example, the mandatory quota set by the European Union requiring 95% of a car to be recycled. Significant technical innovation in product design led to increased efficiency, in the form of better engines with less fuel consumption, as well as improved safety systems. The application of new materials such as carbon and natural fibers continually led to improved sustainability features in cars. Obviously, this would not outbalance the environmental effects through growing numbers of cars and driven distances. Thus, while sustainable car design, efficiency and new fuels continue to be on the radar of innovative automotive companies, two developments were drivers over the past years: the increasing trend towards electric cars and the shift to providing mobility.

TIP

Innovation trends and green mobility opportunities can be found here.

⤳ www.greencarcongress.com

In e-mobility, Tesla challenged the industry. Every major auto manufacturer now offer electric vehicles. Between 2016 and 2018, car companies launched an average of six electric models per year and planned to accelerate that to launch 16 in the subsequent 3-year period. VW, for example, had projected that 30% of its cars would be powered by electric drive trains by 2025. As these projections were made before the corona crisis, we will have to see how the industry develops. But the trend towards electrification will continue. A total of 80% of a car's footprint occurs in the stage of use. While during this stage, e-mobility performs better, its overall footprint is only significantly better under certain circumstances, as battery production does not provide a very good balance. And e-mobility is not the ultimate solution (see page 303).

In the light of sustainability, it is promising to think integrally in offering mobility solutions instead of selling cars. More people, especially in industrialized countries, want to use transport from A to B but do not necessarily want to own a car. When the automotive industry in Germany recently asked for subsidies to stimulate sales in the aftermaths of the corona crisis, a friend commented: *"Cars are so 80's. I want mobility!"* For auto manufacturers, this is a disruptive challenge: to innovate beyond their core product–the car–towards the desire for mobility. This transition is slowly emerging, with cars on-demand in cities around the world, smart solutions to connect different modes of transport and parking solutions, as well as the overall rethinking of urban transport systems. This requires completely new business models, forged by competition between the car manufacturers and new entrants and enabled by technology. In a technology-driven industry, innovation can greatly contribute to transformation. For the multiple challenges, including the aforementioned social ones, there is still much ahead for the industry.

RISKS ON THE RADAR

There is much evidence that policymakers and other stakeholders are now increasingly taking sustainability into account. An interesting development has taken place in the corporate world over the past decade: Sustainability issues have gained in relevance and attention because they have also become business risks.

The World Economic Forum (WEF), a leading and influential organization that, among others, hosts the yearly summit in Davos, publishes an annual "Global Risks Report". This report analyzes risks in five areas: economic, environmental, geopolitical, societal and technological, and surveys a broad range of stakeholders on the relevance and likelihood of issues in these

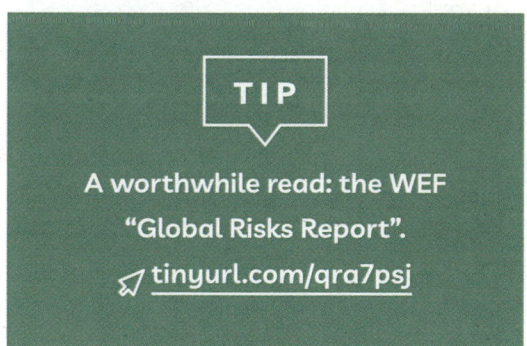

TIP

A worthwhile read: the WEF "Global Risks Report".

tinyurl.com/qra7psj

areas. The annual analysis reveals which topics are relevant—notably from a risk perspective. An interesting picture is emerging over time, as there has been a significant shift in focus. Whereas before 2011, environmental aspects did not even show up in the risk profile, they are now a major consideration, in terms of both their likelihood and their impact. Bear in mind that in 2015, the year of the Paris Agreement, climate change barely made it to the list of top 5 risks. Today, business is impacted increasingly through environmental and social aspects, as well as geopolitical and technological factors that together join the traditional economic risks. It shows that these topics are on company agendas. How awareness translates into relevant action, remains to be seen.

Top 5 global risks over time

2007	2015	2020

In terms of likelihood over the next ten years

2007	2015	2020
Infrastructure breakdown	Interstate conflict	Extreme weather
Chronic disease	Extreme weather	Climate action failure
Oil price shock	Failure of national governance	Natural disasters
China hard landing	State collapse of crisis	Biodiversity loss
Blowup in asset prices	Unemployment	Human-made environmental disasters

In terms of severity of impact at global level

2007	2015	2020
Blowup in asset prices	Water crises	Climate action failure
Deglobalization	Infectious diseases	Weapons of mass destruction
Interstate and civil wars	Weapons of mass destruction	Biodiversity loss
Pandemics	Interstate climate	Extreme weather
Oil price shock	Climate action failure	Water crises

Economic	Enviromental	Geopolitical	Societal	Technological

Based on source: World Economic Forum (WEF). The Global Risks Report 2020. Figure 1. https://tinyurl.com/qra7psj. Accessed December 15, 2020.

ABOUT FORERUNNERS AND LAGGARDS

It is fascinating to investigate and observe sustainability dynamics in different industries, draw conclusions about future developments and derive strategic implications for a business. Sustainability in all its different shapes and forms today impacts corporations and industries, but in different ways and at different paces.

Looking back on almost 20 years of sustainability, I would like to share where I believe different industries stand concerning their level of incorporating sustainability and progressing along a transformational path. At such a high level, this can obviously be only a rough analysis and leaves out the performance of individual companies. The non-exhaustive list primarily serves as a snapshot of the landscape along with some clusters.

– Some industries have dealt for more than 20 years with sustainability issues, especially in the form of reactions to negative incidents. They have implemented advanced mechanisms to monitor and manage certain issues, with each industry having a few leaders bolder than the average. Many complex issues remain to be solved going forward, and the proactive transformation of business models and innovation is still ahead. This is true, among others, for the chemical industry and, to some extent, for the textile and oil and gas industries.

– Several industries have picked up the pace of their transformation over the past years. Some of them react to critical issues, some to opportunities. They have a high impact on their value chains and on other industries, which is why this is a promising development. In this cluster we find, for example, parts of the food industry, retailers, beauty and home care, some of the textile industry, packaging as well as logistics and basic materials, including metals and mining. In transforming these industries, it will be decisive to set the agendas according to the level of challenges ahead.

– Some industries have huge impacts and have dealt for many years with sustainability issues, often as a reaction to public pressure. Nevertheless, they still struggle with it, as sustainability, taken seriously, challenges their businesses model and requires profound transformation. Among these industries are the automotive and energy industries, travel and leisure, parts of the oil and gas industry, the building and construction sector and the banking and insurance industries.

– Some industries could have a huge positive impact on societal transformation but, for different and not always evident reasons, they have still not reached their full potential—neither in terms of performance nor communication. In this group, I would include some pharmaceutical companies, the education and the healthcare sectors, telecommunications, certain technology companies, and some consulting and advisory services.

–Some industries still have a longer way to go on their road to sustainability. Either they have not yet thoroughly started or have initiated small steps but have not yet fully embraced sustainability. Among them are the media industry, the event industry, machinery equipment and other industrial goods, areas of the high-tech industry (software and technology), real estate, consumer goods such as electronics and appliances, luxury goods, and segments of the food industry such as meat, ingredients and beverages.

3.04 —————— NEGOTIATING INTERESTS AND FACING DILEMMAS

Why is progress not being made in so many issues? Let's focus on how sustainability issues advance or cease and how stakeholder dynamics are involved. This will address levels of power and reach. It will also consider competing interests—a common phenomenon in sustainability. There are dilemmas and interests that need to be negotiated.

WHAT DETERMINES POWER?

Stakeholders can have an impact on pushing sustainability. This assertion holds true today for all types of stakeholders: regulatory bodies, consumers, civil rights organizations, shareholders, suppliers, and so on. But in practice, we often don't see things change easily or for the good. So, let's first consider what determines the specific impact of a stakeholder.

Intuitively, power is a major asset. It makes a difference whether I, Anke, want to see change regarding an issue or the German government does. It makes a difference whether Greenpeace engages in a topic or the recently founded initiative of three students from a village in Iceland does. Yes, the example of Greta Thunberg shows: even a single person can start a movement that exerts power on a big scale. Power is not automatically nor exclusively associated with big or highly reputational stakeholders. In part, it remains to be told why some people or organizations gain or are granted power—and this, by the way, is not specific to sustainability.

Typically, the power a respective stakeholder has determines whether a topic will be addressed. A stakeholder with power can influence the direction, as we saw in the example of Greenpeace vs. Shell. Theoretically, this also holds true for other issues. An international fashion brand, for example, could enforce that certain working conditions are in place in all manufacturing sites depending on the level of its power. A leading brand like Ferrero could decide to eliminate palm oil from Nutella and ask their smart people in research and development to replace it while preserving the unique taste. A retailer based on its level of power could require that food suppliers reduce the amount of plastic in packaging. A government could require companies to reduce emissions according to national strategy. They could do these things, by virtue of their power.

Why the power of stakeholders does not "automatically" lead to change has many reasons, often depending on the specific issue. I would like to point out four lines of thought.

– One: Sometimes, an issue is not (yet) pressing enough for a company to change it. Powerful stakeholders don't care enough or don't exert their power. For example, most consumers are generally fine with the status quo. They are not too concerned about palm oil in Nutella, plastic in packaging, precarious working conditions or animal welfare standards. We might want change in our hearts, but we are not truly willing to do what is necessary by adapting our own behavior in the mass market's day-to-day decisions. But if key stakeholders don't care enough, why, then, should corporations care and change? We must not expect companies to be nobler than we are as consumers. Companies carefully evaluate where to invest resources and what is at risk. There must be clear incentives.

– Two: Sometimes, an issue is truly not easy to change. Securing product safety (hygiene, quality, taste, storage life) and reducing the environmental impact of packaging has complexities that are hard to solve. To maintain the taste and consistency of Nutella without palm oil is probably a lengthy, challenging task. And many companies have had the experience that, customers may have a strong reaction if, for some reason, the taste is just slightly different, which leads back to number one again.

– Three: Sometimes, an issue has many different perspectives and trade-offs. Fewer cars to protect the environment implies fewer jobs in the automotive industry and less individual comfort. Higher animal welfare standards or practices such as grazing raise costs and demand more extensive space. The question would be who pays the bill. Consumers, the retailer, the farmer, the producer? Switching to renewable energies implies new infrastructure. But what can we do if nobody wants a wind turbine in their neighborhood? Here, it is less about power and more about negotiating individual interests and finding solutions for complex issues. But, as in all negotiations, no one can have it all.

– Four: Powerful stakeholders often have an interest-based focus on singular issues. In palm oil, the focus has been more on the Orangutan and less on the health of the workers (see page 95). In the case of dairy products, genetically modified feed was prioritized by retailers while deforestation for soy was widely ignored, until now (see page 98). The focus and rigor of powerful stakeholders often decide whether or not an issue advances quickly.

A personal illustration of the last point: Some years ago, I conducted research for a client project on which issues environmental organizations pursued in the textile industry. We identified less than ten relevant organizations, and just a few of them were crucial in terms of their power. Not a single

TIP

The Agrifood Atlas (2017) gives a thorough analysis of problems in global food supply chains. The publishing organizations advocating for change have, until now, had limited power and reachthrough.

tinyurl.com/yx9vw8hn

environmental organization pursued an agenda that covered a broad range of topics, but instead, they basically all advocated for singular issues such as pesticide residues, genetically modified cotton and human rights in manufacturing. As certain aspects were not on the radar of the organizations we monitored, in the specific case, the corporation did not have to fear much pressure. A singular focus on topics can sometimes weaken the power of stakeholders.

MOVING FORWARD, TOGETHER

Over the past decades, a common action undertaken by stakeholders led to progress on critical issues. Let's look at the Roundtable on Sustainable Palm Oil (RSPO) as an example of such a multi-stakeholder approach. The RSPO is an initiative concerned with palm oil—an urgent and still difficult topic with no easy answers but a path is embarked into a good direction.

The RSPO was established in 2004 with the objective of promoting the growth and use of sustainable palm oil in products through credible global standards and multi-stakeholder governance. It currently has over 4,800 members from 96 countries, including growers, consumer goods manufacturers, retailers, banks and environmental organizations.[70]

Palm oil is an ingredient in close to 50% of the packaged products in supermarkets like pizza, chocolate, deodorant, shampoo, toothpaste and lipstick. It has many different properties and functions, which makes it so useful and widely used. Palm oil is also a very efficient crop, producing more oil per land area than any other equivalent vegetable oil crop. Globally, palm oil supplies 35% of the world's vegetable oil demand on 10% of the land. Indonesia and Malaysia make up over 85% of global supply. [71]

Palm oil has been and continues to be a major driver of deforestation in some of the world's most biodiverse forests. The forest loss, coupled with conversion of carbon-rich peat soils, emits millions of tons of greenhouse gases into the atmosphere. It destroys the habitat of endangered species like the Orangutan, pygmy elephant and Sumatran rhino. The exploitation of workers and child labor are problems, as is the disastrous effect burning forests has on the health of local people, as the videos document.

Today, the RSPO represents the largest, independent, third-party standard for the more sustainable production of palm oil, including rules on deforestation, expansion on peat and the use of fire. RSPO certified palm oil is a baseline standard for the mass market.

The World Wildlife Fund (WWF), a major international environmental organization and itself member of RSPO, publishes an annual Palm Oil Buyers Scorecard to critically monitor progress. With the launch of the 2020 Scorecard, a WWF executive stated: *"In the past decade, we've seen companies wake up to the destructive impacts of palm oil production. The biggest brands know they simply can't keep doing business as usual. The risks—both to reputation and supply stability—are too great. Fixing palm oil sourcing is a complex problem. We know what the solutions look like, but adapting and implementing them will take time, innovation, capital, and a willingness to work together. And unfortunately, most companies are still hesitant to take that leap, especially if they think they can do it alone. Platforms like the Roundtable for Sustainable Palm Oil are critical because they convene companies to address sustainability as a pre-competitive issue and drive sustainability across entire sectors. Going forward, we also need to evaluate the impact of different tools businesses can use like long-term contracts to drive sustainability on the ground."* [72]

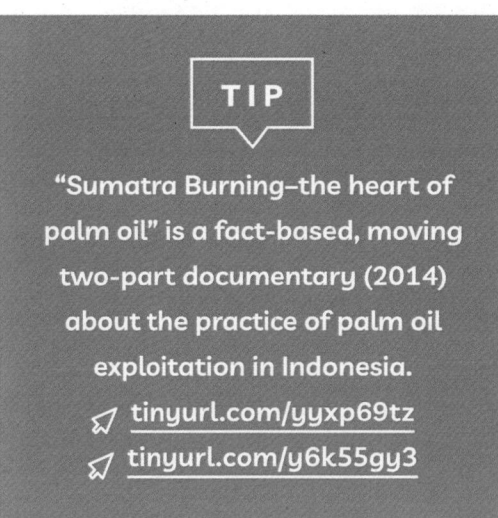

TIP

"Sumatra Burning–the heart of palm oil" is a fact-based, moving two-part documentary (2014) about the practice of palm oil exploitation in Indonesia.

✈ tinyurl.com/yyxp69tz
✈ tinyurl.com/y6k55gy3

Since 2004—and thus, in a relatively short period of time—solutions for a common ground were realized, and many actors could be involved. Over 3,000 companies now apply the RSPO certification standard in their supply chain. Yet, until now, only 19% of worldwide palm oil used is under its regime.[70] The RSPO certification standard grants credibility and reliability and establishes a baseline for a difficult topic. Critics claim, for example, that numerous certified plantations some decades ago were deforested and that enforcement is limited. It is a beginning but not yet the final breakthrough.

WHEN INTERESTS COMPETE

Let's look at a topic that, despite similarities to palm oil, has gained a lot less attention and still lacks a solid solution even though it has been around for more than 30 years. For a client project, I researched deforestation-free supply chains in soy production, analyzing industry supply chains, stakeholders and their power distribution, the evolvement of the topic and the strategic options for the client. To sum it up: the topic reflects the pluralistic appearance of sustainability. The different interests of the stakeholders come up, interact, compete and, in this case, slow down a solution that is good for people and the environment.

Worldwide, soy is grown on an area three times the size of Germany—most of it in Brazil and Argentina. A total of 80% of the soy grown worldwide is fed to pigs, chicken and cows for producing milk, eggs and meat. Forests are basically cleared for soy, palm oil, cocoa, cattle and wood. Statistically, every four seconds, a forest the size of a soccer field continues to be destroyed today, as shown by studies from Greenpeace (2018)[73] and Robin Wood (2018).[74]

> **TIP**
>
> "Are your favorite brands committed to a responsible palm oil future free from deforestation and destruction of nature?" The WWF monitors and reports publicly on companies.
> ✈ tinyurl.com/y6zfj88e

Environmental organizations have been fighting for more than 30 years to stop rainforest destruction. Greenpeace, Robin Wood and Rainforest Action Network, among others, became advocates for the rainforests, especially in Brazil, and the rights of indigenous people. Forests cover about a third of our earth, give a home to 80% of all known species outside of the oceans and provide resources for the livelihoods of 1.6 billion people. Tropical rainforests are especially valuable to our climate balance and biodiversity. Although they cover only 7% of the earth, they inhabit about half of all animals and plants worldwide. The "Soy Moratorium" signed in 2006 theoretically put a limit on the clearing of rainforests. Feed and consumer goods companies, as major buyers of soy, committed to fighting deforestation, as did politicians. A Round Table on Responsible Soy (RTRS) has been in place for years. However, the goal to achieve deforestation-free supply chains by 2020 (which not only applies to soy but also palm oil, wood and paper) has not yet been achieved.

Why? With so many powerful stakeholders committed to the topic and working on it for so many years? Is their commitment half-hearted? Are there too many different standards for soy in the market? Are the controls not strict enough? Is deforestation exaggerated? Are the Southern American governments too lax with putting limitations on production? Is low public awareness a hindrance? Is another enforcement mechanism needed? Most likely, it is a mix of some of these and probably several other factors. Let's take a look at one facet of the broader context and see how the interests of the different stakeholders involved compete.

In our milk, meat and egg value chains—which mainly use soy—deforestation in soy production is not currently a major topic. Even though soy production has been a major concern in the past, the focus was solely on genetical modification (GMO). German retailers enforced it in the dairy supply chain, demanding that dairy farmers employ GMO-free feeding. Within two years, the percentage of dairy products produced without genetically modified soy feed rose in Germany to more than 50%. Protecting rainforests and deforestation-free supply chains, in contrast, have not yet been a focus of retailer's strategies. A few are tentatively starting to address these issues. Some retailers are promoting the growth of soy in Europe. A good approach but not compatible with the mass market. A common approach is missing.

TIP

The WWF UK provides some informative facts about deforestation and a short video game on actions.

⇗ tinyurl.com/y6njxmrn

The most important hotspot for retailers, consumers and advocacy groups in recent years, especially in the value chains of milk and meat, was another topic: animal welfare. Living conditions, grazing, tethered housing and dehorning, among others, have been major concerns, especially in dairy and meat production.

Another hotspot in dairy and meat value chains is climate change, as emissions from agriculture are significant, with approximately 80% of the dairy footprint produced in farming. Higher efficiency—a cow producing more milk through a longer life span or more output per day—can reduce the footprint but might conflict with animal welfare goals. While it is right to address these topics, it implies trade-offs, and a topic like deforestation for soy production is currently left out.

> **TIP**
>
> The burning of Brazilian forests in July 2020 was 30% higher than in the previous year. In August, about 10,000 fires were burning in the Brazilian Amazonas.[75] A complex issue, also fostered by political priorities on increased economic use of the rainforest. See link for more information.
>
> ⊿ tinyurl.com/y3dhlb3y

For some people, the solution might be easy: Vegetarian and vegan diets would solve the problem and stop deforestation. In light of the issues, this is a valid choice. Yet, we should also respect different preferences—such as people's choice of their own diet or the Brazilian government's decisions on how to manage their resources. Other issues also arise, for example, the livelihoods of dairy farmers and their families across Europe and around the world.

From a deforestation for soy production as a raw material, a landscape of issues evolves that present different stakeholder's interests. Sustainability does per se not give a specific value to one issue over another or prioritizes preferences. We, as people, must negotiate across different lines of interests and starting from the question of how we want to live.

Find four issues to reflect on stakeholders and responsibilities. Who is involved? Who must be held accountable? Which actions are necessary for change? Think specifically about companies (which?), politicians (which?) and consumers, plus other stakeholders likely involved.

There is no right or wrong. This exercise invites you to reflect on the complexity of issues and the different perspectives of involved stakeholders.

---

1. ——— Tobacco kills more than 8 million people each year. More than 7 million of those deaths are the result of direct tobacco use, while around 1.2 million are the result of non-smokers being exposed to second-hand smoke.[76]

---

2. ——— Palm oil is used today by basically all major food brands, cosmetics etc. Each person consumes about 8 kg of palm oil per year. The environmental and social effects are serious, among them, deforestation in Malaysia and Indonesia.

3. ——— Worldwide obesity has nearly tripled since 1975. In 2016, more than 1.9 billion adults, 18 years and older, were overweight. Of these, over 650 million were obese. A total of 38 million children under the age of 5 were overweight or obese in 2019. Over 340 million children and adolescents aged 5-19 were overweight or obese in 2016.[77]

4. ——— So-called "planned obsolescence" in industrial design (household appliances, devices etc.) is a widespread practice little publicly discussed: Products are deliberately designed to have a limited life in order to make them unfashionable, no longer functional after a certain period of time or irreparable. This is done to push sales and is largely unprovable and untraceable.

MANA-GING SUSTAIN-ABILITY

In 2019, more than 1,000 top executives across 21 industries and 99 countries were interviewed about the opportunities and challenges to sustainability. The influential corporate leaders agreed that sustainability will be important to the future success of their business and that companies should be making a far greater contribution. The huge gaps between ambition and implementation revealed by the survey reflect my own experience.

I remember numerous discussions about why sustainability management would be needed, whether it differs from regular business, and how it links to existing routines and structures. Based on almost 20 years of management consulting, I've come to the conclusion that companies know how to manage. A key issue is the "where to" and "why" of the sustainability journey. Additionally, a transformation within our given economic framework requires a renewed attitude-driven leadership approach with rigorous scrutinizing and the courage to persevere.

WHAT'S SO SPECIAL ABOUT SUSTAINABILITY?

Let's be clear right upfront: The ultimate goal is to have sustainability harmoniously integrated into regular business. But it takes time until an organization has incorporated and truly absorbed sustainability. During that transformational process, sustainability management is a tool that helps companies to focus and adapt.

TWO ANALOGIES

From a practical perspective, sustainability management identifies and address the challenges and opportunities arising from the sustainability universe. It translates the overall concept of sustainability into specific topics for the company, evaluates and prioritizes them in a strategy and eventually translates them into action points, and where possible, into routines. It implements, interacts with others, measures progress, makes adjustments, and reports on progress. Sounds like business as usual. So, what is new or special?

Comparable to the megatrends such as digitalization and demographic change, with which it interacts, sustainability profoundly affects business and triggers transformation at different levels. It is complex and very dynamic.

But there is another level. This entire book argues that sustainability is a new paradigm for management. It sets a broader vision of business. It is not just about mere money-making but about creating value and contributing to solutions for better lives in our world. In this way, sustainability is transformative and often counterintuitive. It includes a profound shift in attitudes and mindsets towards a broader vision of life and business.

Most companies underestimate how profound sustainability changes their business. And most big and advanced companies overestimate the level to which sustainability is already integrated into the corporation. The

cultural change implanted into sustainability takes time, clear commitments, learning and many concrete steps. I would like to use two analogies
to illustrate the way ahead.

Imagine an old, valuable house that needs rebuilding. Depending on the
architecture and the substance of the house, you will need a few years of
planning, renovating, and modernizing, including installing new electricity, tearing down walls, building up new rooms and so on. A house in bad
shape would be torn down completely and rebuilt on the existing land according to today's standards and desires—a company cannot be torn down.
Instead, it requires a long, continuous process of re-modeling according to
necessities, new rules and possibilities—while you and others continue to
live in the house and the neighborhood.

What has to be done in the rebuilding process is specific to the company
and its business model, industry dynamics, product portfolio and value
chain, while taking into account the business strategy, company structure
and preferences. Consequently, sustainability management has a different
profile in every company.

In the analogy of the house, managing sustainability is not easily or quickly completed. Re-modeling has some heavy implications—complexity,
planning, hard work, delays, difficult coordination. This can seem daun-
ting. But there is a further aspect of sustainability that I want to describe
with another analogy.

It's the analogy of being on a journey. Not a cruise ship or all-inclusive trip
where everything is taken care of. That is not the sustainability journey.
It is more like a backpack trip where you only know your broad destina-
tion. You pack your luggage and buy a good travel guide. You get excited,
looking forward to the new and undiscovered. And the trip unfolds along
the way. The trip's reward comes with the discovery that many of the best
parts of the trip cannot be planned but just received when they happen.
Anyone who has ever taken this kind of backpack trip will agree: it's the
most rewarding way to travel.

To tie this in with sustainability, the picture of a trip brings up and stands
for the broader vision of the journey. I believe we were made to live a
broader vision of business that goes beyond money-making to include

targeting value creation and contributing to a common responsibility. Sustainability provides a worthwhile objective and can lead us to how things are meant to be. To discover and unfold it along the way, is the exciting, smooth and motivating part of the journey.

Sustainability management is a combination of both the fun and discovery of a backpack trip. And it is the serious, methodical and often hard work of reconstructing a house. Along the way, it is sometimes more trip and other times more house.

WHAT GETS US STARTED?

When and how do executives start the journey? The initial impulse can come from many different sides and often from the outside. I would like to make this following passage personal so that you feel individually addressed.

– At an industry association gathering, sustainability is on the agenda. You notice that several of your peers are already on the journey.

– In a management team meeting, an employee explains that rising costs and the scarcity of certain raw materials is putting pressure on a product line.

– Your 9-year old daughter, impacted by Fridays for Future, asks you at the breakfast table what you do in your company to fight climate change.

– You receive a questionnaire from an industry customer, asking about specific environmental data and social compliance inquiries with rules and standards you 've never heard of.

– In social media, you get critical comments on your carbon footprint.

– Your competitor launches a green product range.

– One of your board members brings up sustainability as a topic.

– A new regulation requires you to ban a crucial substance.

– A friend sends you his company's new sustainability report. You flip through it and think: We do all of that, too. I want a report like this.

Several of the impulses have a specific challenge connected with them that needs to be addressed. This is usually a good place to start as it focuses your attention and allows you to learn about sustainability along the way. Instead of planning the whole reconstruction process, you begin by rebuilding a certain area of your house. It may be useful, however, to check with an expert about how this specific challenge interrelates with other likely current or future tasks in your company. Using the analogy of the house, you would not want to start with an area of the house that is not a top priority or one that endangers its entire structure.

Some of the impulses mentioned point to communications, which we will explore in more detail in chapter 5. For now, suffice it to say that communications must match your status quo. To jump into a report or brochure about your merits and the good things you do might look tempting and easy to do. However, the evidence shows that the clear recommendation is: Be careful about starting to paint your house without first making sure that no major renovation or basic reconstruction is necessary. If not, you will likely waste time and money and may damage your valuable reputation. Sustainability reporting is an important step—done well, it can be a great foundation for or give a boost to sustainability. But today, the level of sustainability reporting is high. It is easy to detect from a report whether sustainability is well integrated into a company and where the weaknesses are. Avoid wasting resources, mediocre reporting quality and confusing communications with management.

START WITH KEY PROJECTS

All companies have something in place to start with. For example, they have taken measures to improve energy efficiency or fulfill environmental regulatory requirements. They care for the wellbeing of employees and engage in social causes in the neighborhoods. They make it a routine to consider changing conditions in supply markets. To find out where sustainability adds new pressure or offers opportunities, and at what level in the company things have to advance, consider what is there already. Developing a strategic agenda with focal points and future measures always builds upon existing activities.

A key project—which may arise from one of the mentioned impulses—can kickstart sustainability or drive it to the next level. I define two characteristics for a key project: First, it addresses a complex topic that is critical in its impact on sustainability and on the company. Second, it includes several departments within a company and sometimes also stakeholders from the outside.

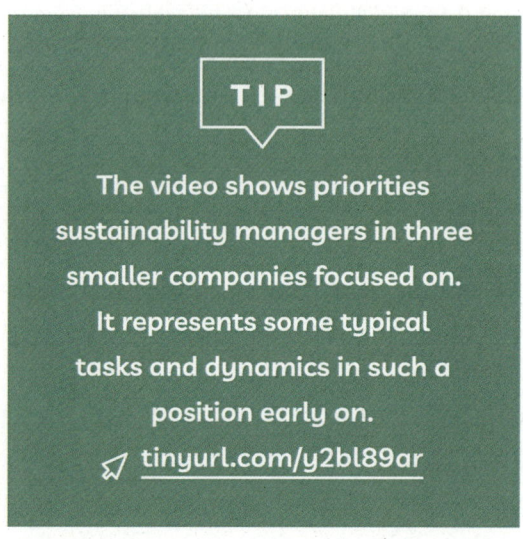

TIP

The video shows priorities sustainability managers in three smaller companies focused on. It represents some typical tasks and dynamics in such a position early on.

⤤ tinyurl.com/y2bl89ar

Key projects have surged often in companies related to supply chain management over the past years. In raw materials sourcing, stakeholders see issues such as human rights and environmental impacts for instance in food supply chains as critical. In copper and other resources processed in industries, such as high-tech and automotive, different issues became critical such as working conditions in mining, child labor allegations and environmental impacts like soil degradation and contamination. Stakeholder's attention and pressure raise business impact.

When supply chain projects surge, they usually affect several departments. In consulting a basic materials company, the affected areas included procurement, business line manufacturing of critical materials, legal, public affairs, environmental management, communications and business development and investor relations—as questionnaires from rating agencies increasingly funneled in. The coordination was handled by the sustainability management officer. First, the status quo was analyzed, including which suppliers, materials, quantities, customers, and contractual aspects were involved. The overall situation regarding regulatory aspects, public opinion, competitor's activities, and the role of industry associations was also assessed. A step-wise plan for adjustments and communication guidelines (e.g., FAQ) were developed. Naturally, not all aspects were solved right away. Instead, the topic remained on the agenda

of the sustainability department where it was monitored for new developments and frequently brought it up in the respective departments. Yet, at some point, they ended up handing it over and, operationally, the new procedures could be anchored at a certain point in the regular procurement practices.

Solving a complex sustainability topic with high business impact at a good pace due to market pressure or opportunities can serve as a blueprint for other topics. It creates an exceptional learning experience for the departments involved and usually uncovers organizational needs for progress, such as missing data or unclear interfaces.

AN EXAMPLE OF GOOD MANAGEMENT

I would like to share a key project case that succeeded in establishing a solution, pushing the company ahead, and extending impact into the market.

In 2007, I advised C&A about introducing organic cotton across its European stores. In 2018, 38% of all of C&A's purchased cotton was organic, making it the world's biggest buyer of organic cotton—a position it had held since 2012.

Worldwide, organically grown cotton has quadrupled since 2005. In 2018, the global production of organic cotton increased 56%, and this growth is expected to continue. Still, only 0.7% of total global cotton production is currently organic.[78]

Conventional production practices for cotton involve the massive application of substantial fertilizers and pesticides, the extensive use of water, soil erosion, soil and water contamination, and health issues for farmworkers.

Organic cotton has a significantly better eco footprint. For the amount of organic cotton instead of conventional cotton C&A bought in 2017, it saved 170.8 billion liters of water, avoided the use of 157 metric tons of hazardous pesticides, and improved the quality of over 174,000 hectares of soil.[79]

C&A had been on the sustainability journey for several years, especially in terms of social compliance and working conditions in factories mainly in Asia. In 2005, they started to purchase organic cotton. In 2007, a carefully planned market entry into retail stores became a milestone for sustainability. Let me share seven things C&A did well when introducing organic cotton.

1. Choosing the right topic at the right time. Organic cotton is a topic highly relevant for environmental and social reasons and one that C&A could influence. C&A took on the topic very early; it was a niche topic not yet addressed in the mass market. As an early mover, they could shape it and build up supply chains that ended up beneficial later on.

2. Asking for help. They decided to work with a consultant to help them navigate the process. We analyzed stakeholders, outlined C&A's unique selling points, and developed a step-by-step communication plan for the pilot and rollout phases of organic cotton.

3. Getting the business case right. C&A is a textile retailer for the mass market with price-sensitive clientele. The goal to get organic cotton into the mass market would not have worked using a typical strategic approach of making it more expensive than conventional cotton textiles. C&A took a wise and bold decision: they introduced organic cotton at the same price as conventional cotton and bore the cost for the surcharge on their own. Later on, when the market for organic cotton gained momentum, it became a major success factor because C&A had already set up its supply chains early on.

4. Understanding and then managing the playing field. C&A was highly aware of the potential risks and the necessity to benchmark, leading to careful analysis and foresighted management. It collaborated with organizations such as the Textile Exchange and the Organic Cotton Accelerator, as well as with local partners in China and India through the C&A Foundation.

5. Starting with a pilot phase and rolling out with determination.
C&A piloted organic cotton into numerous branches in Germany, choosing a relevant, large enough market and not only planning but also implementing a clear rollout strategy.

6. Communicating honestly and in a targeted way. Communication was fact-based and sympathetic, focusing on the positive environmental effects of organic cotton and the commitment to make it available at no extra cost. A C&A organic cotton label was developed for products and store hangers.

7. Using the momentum to leap ahead. The success in organic cotton triggered the next steps. In 2012, the "we CAre" sustainability strategy was released. In 2018, C&A invited customers to join the journey with a broad portfolio of sustainable products under the label #WearTheChange.

In the years since the launch of organic cotton, C&A has faced several challenges in business and sustainability—but still remained well on its journey. For more information, check out the C&A website and its sustainability report.

⇗ www.c-a.com/sustainability-report   ⇗ tinyurl.com/y4pqpohv

⇗ tinyurl.com/y59k5f3v

STRATEGY: DECIDING ON ACTION

A sustainability strategy defines a few focal points a company should strive to pursue. Ideally, it defines clear targets and action plans for these focal points over a certain period. In larger corporations, a materiality analysis should lay the foundation for the sustainability strategy.

Companies are also increasingly considering the Sustainability Development Goals (SDGs, see pages 52f.) and connecting them to their strategic focal points. According to a recent PWC report, 50% of companies do so. 71% of CEOs believe that—with increased commitment and action—business can play a critical role in contributing to the Global Goals, as the SDGs are often called.[24]

There are typically two approaches: Deriving a strategy from the SDGs or linking the existing sustainability strategy to the SDG global agenda. From my experience, the second approach is prevalent. Moreover, many companies do not apply the SDGs in the way they were intended. Some companies merely combine existing activities and goals and "relabel" them to fit into a recognized framework rather than setting new goals. Yet, some broader considerations may be necessary in order to define which significant contributions would qualify for the SDGs until 2030. In a recent client consultation, we jointly decided to postpone the display of the SDG contributions. Due to other priorities, the company is not yet at a stage where it can define strategic contributions to the SDGs. We had not only wanted to label the existing activities with an SDG link but to firmly commit to them.

Let's highlight some general strategy aspects based on examples from different companies.

The consumer goods multinational Unilever is an example of how a straightforward strategy can be designed despite reaching across a complex business. The "Sustainable Living Plan" defines three main focal points—health and wellbeing, environmental footprint, livelihoods. An overarching target is set for each focal point that is further translated into an operational goal. Selected topics from the materiality matrix, such as Health & hygiene or Packaging & waste, are connected to these goals (see page 65). The SDGs are linked to the strategy. Progress reporting is at a sophisticated level as can be seen on Unilever's website: tinyurl.com/yynq2w4x

Aurubis is a leading worldwide provider of non-ferrous metals. The Hamburg-based company processes complex metal concentrates and diverse recycling raw materials and belongs to the global leaders for copper recycling. In the face of rising demand for metals due to digitalization and urbanization, as well as the growth of e-mobility and renewable energies, Aurubis contributes to the supply of raw materials. Responsibility is one of the three cornerstones of Aurubis' business strategy. Within this, the sustainability strategy until 2023 comprises nine areas of commitment

The Unilever Sustainability Plan

We have three big goals

Improving health and well-being for more than 1 billion	Reducing environmental impact by half	Enhancing livelihoods for millions
By 2020 we will help more than a billion people take action to improve their health and well-being.	By 2030 our goal is to halve the environmental footprint of the making and use of our products as we grow our business.*	By 2020 we will enhance the livelihoods of millions of people as we grow our business.
Health & hygiene Improving nutrition	Greenhouse gases Water use Waste & packaging Sustainable sourcing	Fairness in the workplace Opportunities for women Inclusive business

Based on source: Unilever website. tinyurl. com/yynq2w4x. Accessed November 28 2020.
Slightly adapted with permission of Unilever.

that balance environmental, economic and social key topics. In each of the nine areas of action, a strategic approach, a target, key measures and key performance indicators (KPIs) are defined. This refined, detailed breakdown helps to manage well in daily business.

As an example, see the breakdown for "Recycling solutions", one of the three economic key points and very specific to Aurubis' business model. Aurubis is one of the leading recycling companies for e-waste, a major sustainability issue (see pages 284ff.).

More detailed information on the strategy and the other areas of action can be found on the company's website. Here you can also find the company's sustainability reports, yearly KPI updates and the Non-Financial Report available to download. ✐ tinyurl.com/yykv4q5y

Recycling solutions

Strategic approach

We invest in our multi-metal recycling and, in this way, contribute to a circular economy and thus to the conservation of natural resources beyond our key expertise in copper recycling, with excellent energy effciency.

Target

We have set the target of using a larger volume of complex secondary raw materials in addition to copper raw materials, extracting many metals apart from copper and making these metals useful for society.

Key measures

» Increasing the volume of complex recycling materials sourced
» Establishing and developing "closing the loop" systems as a result of new or intensified cooperation with original equipment manufacturers (OEMs), retailers, or copper product customers
» Analyzing market conditions and future opportunities for sustainable products (think tank for metals, products, and services)

KPIs

» Direct sourcing of complex recycling materials from collection points: 100 % volume growth by FY 2022/23 (base year: 2016/17)
» Number of "closing the loop" systems with direct and indirect product customers from the metal value chain: increase in target by 10 by FY 2022/23 (base year 2017/18)

Source: Aurubis website. tinyurl.com/yykv4q5y. Accessed December 7, 2020. With permission of Aurubis.

4.02 ——————— MAKE IT WORK— A LOOK INTO THE TOOLBOX

Let's continue to look very practically at installing sustainability in a company. While it depends on the specific case and situation, some instruments are in most toolboxes. Please also refer to additional instruments in other chapters such as materiality analysis (pages 62f.), stakeholder analysis (pages 75f.), footprints (page 58) and report (page 158).

SETTING RULES

Guidelines and policies are commonly applied to a variety of company topics for rule setting, and therefore also cover several sustainability topics.

Guidelines seek to simplify a set of processes for an established habit or practice and codify it. Although not mandatory, guidelines ideally impact an entire system in a positive way. Policies, in contrast to guidelines, provide a general framework for a certain topic and are typically compulsory.

A code of conduct is a guideline that is widely applied. A code of conduct for employees sets the rules, among others things, for how a company expects its employees to behave in reference to values such as fairness, respect and honesty, to rules to prevent racism and discrimination, and terms of environmental behavior, for example, the conscious use of resources such as energy and water. A supplier code of conduct does the same by establishing the basis for interaction between a company and its suppliers. Here, for example, a company specifies what it expects from suppliers in terms of human rights, fair wages and protecting the environment. Other examples of guidelines include rules for plant health and safety or guidelines for marketing products responsibly to children.

Management of sustainability often includes a broad range of policies that codify rules for either internal conduct or interaction with stakeholders. They guide a company in attaining positive results. An environmental policy, for example, includes a high-level commitment to different aspects of environmental protection. Such a policy is a precondition for an environmental management system such as ISO 14001.

A sourcing policy sets the requirements and expectations a company has towards its suppliers—often applied today for instance in the food industry. The Unilever Responsible Sourcing Policy (RSP) enforces the company's strategy. It is instrumental in ensuring that *"Unilever meets its business objectives while making a positive social impact on the lives of millions of people in supply chains around the world and reducing the environmental impact"*. It is complemented by the company's own Sustainable Agriculture Code (SAC), which recognizes external certifications for certain products such as palm oil, soy or cocoa.

Guidelines and policies are good instruments to structure priorities, lay cornerstones, and provide orientation and a basis for implementation. They don't automatically enforce change and usually achieve only the level of compliance.

Set requirements, expectations and an aspired level of commitment serve only their intended purpose when applied and enforced. This often poses challenges when dealing with complex issues such as human rights and environmental protection in supply chains. Thus, audits are a typical additional measure applied to monitor and enforce guidelines and policies. They can be conducted internally or as external, independent, so-called third-party audits.

The practice of auditing raises the issue of multiple audits. Imagine a supplier is connected to numerous customers—which is common in many industries such as food, textile and automotive. If each customer were to audit each supplier, it would be a tremendous waste of time and money. But competition-sensitive data about supplier-customer relationships cannot be easily exchanged. SMETA (Sedex Members Ethical Trade Audit) offers for example an efficient and cost-effective digital solution. Using this platform, a company is able to share its audit results with multiple customers, and data sets can be matched without revealing sensitive data. Sedex is one of the world's leading ethical trade service providers that strives to improve working conditions in global supply chains with the four pillars Labor, Health and Safety, Environment and Business Ethics. Nearly 20,000 SMETA audits are uploaded onto the Sedex platform every year.

INTEGRATING INTO EXISTING SYSTEMS

Management systems are in place in most companies. Sustainability is therefore either integrated into the existing management systems or explicit sustainability management systems are installed. For example, environmental, health and safety (EHS) management systems operate and monitor EHS measures across many multi-site corporations with a manufacturing focus. Among these, for example, are companies in the oil, gas, metals and mining industries. Management systems monitor inputs (e.g., energy used) and outputs (e.g., accidents) along key performance indicators. In the food industry, quality and product safety management systems often integrate environmental and occupational health systems. Aspects of social and environmental performance are also increasingly integrated into supply chain management.

Environmental management (ISO 14001 or EMAS), energy management (ISO 50001), and occupational health and safety (ISO 45001) are examples of typically applied internationally certified management systems with a dedicated focus on sustainability topics. They support management, foster improvement, and are a signal of commitment. As certified systems are associated with bureaucracy and costs, they are usually applied when an economic benefit is expected. In some industries, for instance, it is standard to expect manufacturing sites to be ISO 14001 certified, which then may become a baseline.

LABELS: SIGNS FOR ACTION

Standards and certifications are also very widespread and commonly applied to products, usually by transforming them into labels. There are certifications for specific topics such as Fairtrade®, which stands for surplus wages for farmers and covers coffee, tea, chocolate, fruits and textiles, among others. FSC® is for sustainable forest products, MSC® and ASC® for sustainability in seafood (see page 295), and GOTS for organic textiles. Some product labels have a limited regional scope such as Energy Star®, certifying products for energy efficiency in the US, and "Blauer Engel", certifying a range of eco products in Germany.

Some certifications cover specific industries and thus have their criteria targeted to the industry-specific sustainability challenges. These are not always the highest standards in terms of sustainability performance, but at least they set the bar higher. Let's take a look at three examples.

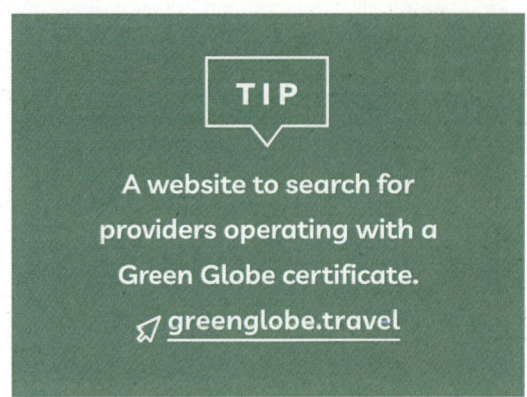

TIP

A website to search for providers operating with a Green Globe certificate.

🚀 greenglobe.travel

– The international Green Globe certification for the tourism industry covers beachside resorts, urban hotels, congress centers, rainforest eco-lodges and so on. Green Globe certifies the positive environmental and social activities of members and targets continuous improvement. It can be seen as a baseline standard that triggers activities in an industry with huge negative impacts.

– The Charter for Sustainable Cleaning is a voluntary initiative set up in 2005 for the European soap, detergent, and maintenance products industry. The logo is on-pack of many products and reflects that the product meets the specific sustainability criteria agreed upon in the industry. It is also a baseline standard.

– LEED (Leadership in Energy and Environmental Design) is a green building certification that includes the design, construction, operation and maintenance of buildings, homes, and neighborhoods to ensure they are environmentally responsible and use resources efficiently. Another standard in this area is BREEAM (Building Research Establishment Environmental Assessment Method). Published in 1990, it is the world's longest established method of assessing, rating, and certifying the sustainability of buildings. More than 550,000 buildings have been BREEAM-certified, and over 2 million are registered for certification in more than 50 countries worldwide. Both are relatively high standards.

Standards and certifications aim to align performance with clear criteria and thus, provide comparability. To decide which instrument is appropriate, comparisons need to be done that consider the purpose, costs and benefits.

TIP

The website lists labels and standards for different industries and countries. They promote a "World Ecolabel Day" on Oct 8.

🚀 www.globalecolabelling.net

INDICATORS FOR CHANGE

Indicators, often called KPIs (key performance indicators), are an important instrument in sustainability management. As such, they are an integral part of the management systems and at the heart of reporting.

Indicators make the status quo and change tangible, allow for comparisons between companies and over time, and are important for detecting key leverage points. Indicators themselves don´t improve sustainability; they need to be linked to targets and measures. Due to their relevant but broad and complex nature, we will focus on only a few major insights and examples.

TOOLBOX

THREE RECOMMENDATIONS FOR HANDLING INDICATORS EFFICIENTLY

1. First improve your housekeeping: Many indicators today are mandatory for reporting on ESG, GRI, CDP or customer surveys. Focus on a good setup and invest in systems and processes to minimize effort in handling and monitoring the data you need to report on a regular base.

2. Align your indicators where possible to international standards: Indicators sometimes seem "home-made" and lack transparency in terms of methodology and confirmability. Wherever possible, align indicators to available standards. An example is to use footprint calculations available for certain industries and products as reference scheme for your own footprint. Another example is to define climate action science-based targets that are aligned with the Paris Agreement´s protocol.

3. Define a few indicators specific to your business and your strategy: When companies advance in sustainability, their uniqueness is not decided by indicators on emissions or diversity

The following examples illustrate good approaches for such indicators specific to the business and strategy.

Aforementioned Evonik measures how its business impacts society. Three indicators are calculated and displayed by the company, among them the following: "1 : 7.9 jobs". One Evonik employee secures an average of 7.9 jobs in the value chain. Graph and methodology can be viewed at the Evonik website. ⚐ tinyurl.com/yxlppwgy

BMW annually tracks the development of resource and emission intensity per vehicle produced and displays this indicator in its Sustainable Value Report. ⚐ tinyurl.com/y85sv6nm

Change in resource consumption and emissions per vehicle produced compared with the previous year	2017	2018	2019
	-5.3%	2.7%	-7.8%

Organic coffee, tea and spice producer Lebensbaum unusually discloses what portion of the price a consumer pays for a bag of coffee goes to the farmers: it´s 26%. 24% of the price goes to Lebensbaum, 20% is used to pay taxes, licenses etc., and 30% is spent on distribution. By making the price breakdown transparent, the company reveals its intention to care for farmers and pay decent wages. The graph "Price transparency—a look inside the bag" can be found in Lebensbaum's Sustainability Report.
⚐ tinyurl.com/y5qf4kyh

Unilever displays how it performs related to a target set in its strategy.

Strategic indicator at Unilever

Educating people around the world on good oral health

An estimated 3 billion people don't brush their teeth twice a day. Most use less than half the toothpaste dentists recommend for brushing properly. And 1 billion or more don't use fluoride toothpaste at all.

That's a lot of people we can help – and a big prospective market for our oral health brands.

Worldwide, we've reached a total of 102 million people through our Brush Day & Night educational programme, free dental checks and TV adverts promoting the importance of oral health and hygiene – vastly exceeding our target of 50 million by 2020.

>102 million

By 2019, our Smile brand had reached a total of over 102 million people, more than double our 2020 target

Source: Unilever website. https://tinyurl.com/yy5fvngt. Accessed November 28, 2020. With permission of Unilever.

4.03 —————— SET INCENTIVES FOR SUCCESS

TIP

Corporate Knights has a YouTube Channel with examples on "clean capitalism"
tinyurl.com/y36xg52o

and publishes an annual list of company performance.
tinyurl.com/y6x4ufv2

Incentives are an instrument widely used in companies, especially performance-related financial rewards such as bonuses or special benefits, which include work flexibility, training and development support and job promotions. A basic reason to use incentives is to achieve business goals and motivate people in their work. Which role do incentives play in sustainability, and how does their use differ from familiar practices?

LINKING INCENTIVES TO PERFORMANCE

Investors increasingly ask that sustainability should be linked to incentives, for example via ESG criteria (see pages 55f.). They call for reward schemes, especially for executive bonuses that are tied to one or more sustainability-related performance targets. This is already common in heavy industries (e.g., utilities, materials) where sustainability issues such as climate change, water availability, pollution and safety are meanwhile understood to have material impacts on the company and, in many cases, are regulated.[80] Also, among leading sustainability companies, such as the ones presented in the "Corporate Knight's" Report on the global most sustainable corporations, a large number provide a monetary bonus to executives who achieve sustainability targets, among them Adidas, BMW and Schneider Electric. Overall, however, this instrument is not yet widely applied, as research for the UK shows.[81]

TIP

Two Harvard Business Review articles illustrate how to link sustainability to financial renumeration of executives.

🖅 tinyurl.com/y3gscxl8
🖅 tinyurl.com/y3zknpn8

Setting incentives was one of the four measures suggested by McKinsey to bring value to sustainability programs. The top reason that survey respondents gave for their company's failure to capture the full value of sustainability was the lack of incentives to do so. Among survey respondents, 37% named short-term earnings pressure as a reason for poor sustainability results; about a third named lack of key performance indicators and not enough people being held accountable.[82]

While incentives will not "fix it all", their importance will rise over the next decades. For a simple reason: It is the logic of businesses, where the common understanding is to link financial rewards to performance. This same mechanism falls short when applied to performance with respect to sustainability. When designing incentives that will truly foster sustainability performance, from my perspective, two aspects need to be addressed: an organizational and a motivational.

WHAT ABOUT ORGANIZATIONAL READINESS?

Research on incentives points out several reasons why sustainability-related incentives are either not installed or not functioning well. Some of the challenges are organizational and reflect my own experience.

– ESG performance-related topics are usually under the responsibility of middle management. At this level, incentives are less often linked to financial performance. Linking ESG to the executive bonus would not solve the problem and not set adequate incentives.

– Sustainability in most companies is still addressed in a silo. While there may be one executive who is responsible for sustainability performance, it is still not considered a cross-organizational or broad management priority. This is a widespread overall challenge that cannot be solved with incentives. Installing incentives in such a situation with a lack of organizational readiness would very likely cement this status quo and not achieve any improvement.

– Not all relevant sustainability topics have a clear set target or improvements that can be directly quantified. This general challenge cannot be solved by establishing financial rewards.

– Overall, there is little transparency on how companies link sustainability to compensation. Details on issues prioritized, weight assigned to sustainability factors and where this is applied are rarely publicly disclosed, not even to investors. As a result, it does not fully realize its value.

In summary: A careful check needs to be done on which employees receive a bonus and related to which criteria, to truly unfold its impact. In today's setting and sustainability's overall status, financial incentives do not work well.

WHEN FINANCIAL INCENTIVES FALL SHORT

It is interesting that money is brought into the game for incentivizing sustainability. This is typical business logic. I would like to challenge it with a counterintuitive perspective related to sustainability that I have observed in practice but which has, so far, been widely neglected.

A powerful approach—maybe even more relevant than money—links immaterial incentives to sustainability, namely freedom to innovate, dedicated resources and attention. What do I mean by this?

An involvement in sustainability comes from the heart for many people. It gives them a purpose (see chapter 7). These employees—many of them young—want to make the world a better place, contribute to saving the environment and build a fairer society. They want to see their company do so and have an impact themselves by participating. Money is not a primary motivator. Usually, these people earn a good salary. A financial incentive does not necessarily motivate them. Not to say that money never works as an incentive for some of these employees. But I propose three measures for immaterial incentives. The potential of these measures can unlock motivation and may even lead to better sustainability results.

TOOLBOX

THREE ALTERNATIVE WAYS TO SET SUSTAINABILITY INCENTIVES

1. Freedom to innovate: Grant people time to develop ideas to improve sustainability—either proactively or when they approach you. For example, by allowing a certain amount of time to be dedicated explicitly to sustainable ideas or offering corporate challenges associated with sustainability. When employees have ideas on sustainability, take them seriously. A proactive way would be to invite employees once a year to a workshop about imaginatively visualizing how your company could address certain societal challenges (see page 184). Invite a mix of employees in terms of their level, experience, and departments.

2. Dedicated resources: Grant resources to employees to drive selected projects over a certain time. Most initiatives in sustainability, especially in the early stages or with regards to future topics, notoriously lack time, money, staff and the courage to be pursued properly.

People will value the trust and space they are given and will surely fill it with great outcomes.

3. Grant management attention: A steering group, a "godfather" for a major project, a sincere interest in the progress of major sustainability initiatives, a contest with a management jury, mentioning employees in company meetings, a quarterly or monthly jour fixe, or 15 minutes at each of your management team meetings dedicated to sustainability. All of this will mean so much more to the employees in charge than an extra paycheck. Why? Because they feel validated, which makes the chances for the initiative's implementation rise.

4.04 —————— ON THE JOURNEY

Determining key projects and establishing a strategy are tremendous milestones. But what if you don't have the structure or organizational set-up in place to do this? What if there is no obvious key project out there, or your company is not yet ready for a strategy? Or, as an advanced company, you may even have a strategy but are stuck in getting big changes implemented. According to the 2019 CEO Survey, business execution is not measuring up to the size of the challenge, for instance in climate change:

"Broader commitment on science-based climate action has actually declined since the Paris Agreement in 2015: this year only 35% of CEOs say they have or plan to set a science-based target in the next year, compared with 43% of CEOs that told us reduction targets in line with science was one of the most important climate leadership behaviors for companies to adopt in 2015 ... In many cases, keystone actions, which are challenging for companies to implement, are being postponed. While just a third of CEOs are willing to commit to a science-based target in the next year, more CEOs (44%) are comfortable with the statement "a net-zero future is on the horizon for my company in the next ten years."[24]

How can the journey be started, and how should delivery be executed? Let's look at the greater plan for rebuilding your house and the pathway of your journey. But first, let´s look at the necessary organizational setup.

FINDING MR. OR MRS. RIGHT

The presence of a sustainability department or person is mentioned repeatedly. Such departments and persons play a crucial role in advancing sustainability in an organization.

From my understanding, a sustainability manager (or team) is ideally a focal point for sustainability: a knowledge center, networker, advisor, integrator, guard, developer, administrator, communicator, challenger and solution facilitator all rolled into one.

I would like to take a moment and expand on this point because, in my experience, one of the most understated actions to take is to install someone who is a good fit as a sustainability manager within a company. What does this mean on a practical level? Who would be a good fit as Mr. or Mrs. Sustainability? And where should she or he be put in your organigram?

First, sustainability management´s home base is diverse. I know of sustainability managers in all different departments: communications, environmental affairs management, public affairs, strategy, product quality, procurement, supply chain management, human resources, investor relations, business departments, marketing & sales, research and development, and the CEO office. But not, for example, in finance, legal, operations, IT or administration departments. Frequently, a specific position or department is installed as a staff function with direct reporting to the management team. Often, a starting point is the department where a key project first emerged (see page 108).

Secondly, it is not decisive if the person responsible is young or old and more experienced, a man or a woman. However, it does appear that more women than men are in positions associated with sustainability. Maybe this is the case because it is truly a multi-task job, which women are supposedly a better fit for.

 **TOOLBOX**

SEVEN POINTS FOR SELECTING A SUSTAINABILITY MANAGER

1. This is a tough job. Fill it with a person who will be respected in your company. Give them all of the backup they need to take on the role and execute it well.

2. Choose a person that has a firm positive attitude on sustainability but also a good sense of business priorities. The latter can be more easily developed than the first. It should be a priority to develop skills accordingly.

3. The importance you award the position will reflect on sustainability.

4. Choose a department that has a true influence on daily routines and strategic decision-making.

5. Provide a position. Most sustainability managers start with their assignments on top of their regular jobs and fight for years for a standalone position.

6. Give a clear job description. Most companies are lost with this task at the beginning. Get help.

7. Grant management attention. Actively demonstrate that you care—through time, budgets and awareness.

What these points mean in terms of selection criteria, job level, structure, persona and so on, must be decided and designed for the specific context. Let me underpin this with two very distinct examples of how two client companies got started—it is a coincidence that both sustainability managers were women.

For a food brand, a very young graduate with a high passion for sustainability was hired. This person had a certain level of sustainability knowledge from university. Two experienced people from related departments (sourcing, marketing) became chief executives and mentors. The person worked on several projects focused on operations and communications, developed guidelines and advised departments on issues such as packaging. I was then hired as a consultant for the strategy process.

In a high-tech company, a woman in her late 50s with extensive experience in business functions and finance, took on the sustainability job. Her vast network, business knowledge and level allowed for a relatively quick and smooth process of integrating sustainability into the business in the first wave. The knowledge in sustainability that was lacking was brought in from the outside. Over the period of a few months, I onboarded her and explained all she needed to know about sustainability, and together we led a targeted strategy process.

Make sure that, at a minimum, you start with a dedicated person who has the openness to expand to a small team. The structure needs to be reviewed frequently and adjusted over time. New tasks and roles must be defined with the growth in staff. The key aim should be, however, to eventually dissolve the team and have sustainability deeply anchored within the company.

TEN TASKS IN THREE STEPS

To implement sustainability gradually in the corporation, let's look at the full picture and break it down into ten tasks in three steps over the course of several years. Time frames are highly individual but are usually longer than originally envisioned as topics multiply along the way. The main pitfalls are a summary of the challenges faced when advancing to the next level.

This "unisize" plan for a company with global supply chains and multiple sites needs to be further detailed within a specific context. Smaller companies can downsize it to the main impact points. As a corporate leader, be invited to assess where you are and cross-check with the detailed descriptions.

Step	Major tasks	Approximate time frame	Main pitfalls
I. Get ready	1. Evaluate the status quo 2. Start an action plan	6 months to 2 years	Little dedication to the process, lack of know-how, over- or under-analysis
II. Take-off	3. Set up monitoring 4. Engage with stakeholders 5. Prepare for reporting 6. Develop a strategy	About 3 to 8 years	Lack of organizational embedding, over-ambition, too little perseverance
III. Advance	7. Implement 8. Take it to the next strategy level 9. Address new key entities 10. Overcome organizational barriers	At least 5 years	Lack of management attention, lack of know-how, pride, inside-focused, biased

Own presentation

A review of the status quo that cross-checks the outside view with the inside perception can be useful, especially for companies that have some years of experience in sustainability and are stuck, as many corporate leaders in the CEO survey for example stated. These advanced companies are from my experience often on the threshold between take-off and advance.

Here below is a description of the steps and major tasks.

Step I: Get ready

1. Evaluate the status quo

An internal inventory of the current activities in sustainability is conducted: What has been done and by whom and with what results? Which guidelines, systems, visions, strategies, targets, indicators and people are in place? Are there focus areas? What are we good at, where could we improve? What is the motivation for change? Collect information on the status quo and structure it. An approved framework would be strategy, communication, actions, structures and processes. Additional layers such as the products, the supply chain and the people could be added.

The outside perspective should be considered in further analysis. Outside-in competitor benchmarking is conducted. Major stakeholders are listed and their concerns analyzed. The impact of mega trends on the business is analyzed. Industry-specific topics are collected. The topic list needs to be evaluated, bundled and clustered, for example, into risk, neutral, and opportunity. A materiality analysis at this point is not yet necessary. The task here is to understand and structure the major trends and topics in the company's environment and benchmark its own performance and organizational readiness for the most relevant topics.

2. Start an action plan

After analyzing the status quo, the recommended priorities are derived for an action plan. This action plan can range from 6 months to 2 years. All of the results should be discussed with the management board to gain a common level of understanding and support for the action plan.

Existing frameworks (GRI, SDGs, UNGC principles, ISO 26000, DNK etc.) can give guidance but should be evaluated for purpose and their useful application. More importantly in this step, however, is to assign a person to be responsible and conduct a solid, insightful analysis that leads to relevant action.

Step II: Take-off

3. Set up monitoring

With an action plan in place and, likely, key projects running in parallel, the next major task is to prepare the business for increased transparency and steering. Many companies at some point start to receive inquiries from customers (especially business-to-business companies) or from investors and rating agencies (for stock-market listed companies). The company needs to become knowledgeable in case of inquiries and to detect gaps within the organization. The data necessary depends on the specific setting but usually includes data on the information requested by stakeholders and a range of indicators (see page 119).

The set-up process includes collecting numbers, defining the scope of data, aligning the data available, setting up processes for regular data collection and monitoring. At this stage, all of this will call for tremendous focus and pragmatism. At the latest now, a dedicated person should be assigned to this process.

4. Engage with stakeholders

Now would be a good time to move to the next level of openness. Communication can start, for example, with a short overview, FAQs and factful statements on issues of public concern (see page 155). Engaging with stakeholders to gather knowledge, build networks and become more visible as a company could include participating in multi-stakeholder groups or working groups at an industry level, or presenting the company at panels or universities. A concerted approach within the company is now needed, as usually many different departments and people at this point will already have some interaction with stakeholders. It is useful to design the approach as an educational journey, to receive input and feedback on relevant topics and build trust. The engagement should be intentional, carefully evaluating input and output.

5. Prepare for reporting

You either want to or need to report on sustainability. Probably, you already have some reporting in place. The specific reporting requirements depend largely on the status quo and the goals (see page 158). It is at this point that validated stakeholder and materiality analyses should be concluded. Sustainability reporting is prepared, sketched out and orchestrated across a defined timeline. A content-based, mid-term plan for communications across different channels should accompany the reporting process.

From my experience, a balanced approach involving taking action and proper reporting is needed. Many companies start reporting before they have things well in order. Communicating on the fly is no problem as long as the awareness and action for a parallel process of communication and management is in place.

6. Develop a strategy

The definition of focal points, a strategic plan with targets, key performance indicators and measures were described earlier. Ideally, the sustainability strategy should include all insights from the analysis of the previous steps and be aligned to the business strategy.

Some people will argue that number 6 is a prerequisite for number 5. Yes and no. Technically, a requirement in reporting is to have a strategy. From my experience, during the first years of reporting and management, most companies do not yet reach the level of a full-scale sustainability strategy with targets. The most they achieve is a strategic approach with some rough focal points and management approaches and qualitative goals that often still stem from a communication perspective. As argued, this is no problem as long as there are awareness and advancement. A strategic review, including benchmarking with peers and recommendations based on the materiality analysis, will help to develop a strategy with targets at a certain point in time.

Based on my experience, the threshold for the next phase can be described as follows: A successful company has set up a clearly defined process with key actors from within the company for strategy definition and can count on knowledgeable top management involvement. There is an internal awareness of some major organizational barriers, which have usually come to light by this point. Over the course of time, the company has experienced pushback and failures and has had to exercise perseverance. A well-functioning personnel sustainability infrastructure is in place, and the topic is brought to a new level of attention and embedded in the organization, which is necessary for implementing a sustainability strategy and transitioning to the next stage.

Step III: Advance

7. Implement
The implementation of the strategy with associated actions—such as aligning processes, extended reporting and putting the measures into action, will take some time. Key entities within the company (certain departments, sites, people) have to be integrated. External stakeholders (such as suppliers or community partners) are often involved in executing the strategy.

At this point, the sustainability department is still in the role of orchestrating the overall process. It advises, aligns and monitors activities, supports the installment of instruments such as policies, management systems, standards and incentives. Yet, it needs broader organizational integration and pull from the top but also middle management. This is a crucial point to spread sustainability broader throughout the organization. Many companies miss this by not getting employees on board at a broad level and not intentionally addressing internal key stakeholders who could act as multipliers. Also, at this point, sustainability has to be integrated into the overall management processes (incentives, reporting routines, innovation priorities, regular communications etc.). It is decisive to come out of the sustainability silo and take a big leap into integration. In my experience, many companies are stuck in just that major task.

8. Take it to the next strategy level
There is no way to predict or plan when a company will reach this point. It depends on a mix of several factors coming together in the right place at the right time. Certainly, external pressure, such as ambitious pledges or industry disruption, is an excellent motivation to leap ahead. Top management commitment can also serve as a pull.

Diving deeper into topics, meeting ambitious targets and innovating beyond current business routines requires new mindsets, a new level of commitment and, accordingly, new structures and processes. Executing a climate strategy with science-based targets in accordance with the Paris Agreement, for example, requires an immense, united effort throughout the organization. Pushing innovation might require making it a strategic priority. Bundling projects into a lighthouse project or portfolio will increase impact and awareness. A major challenge is to define and execute sustainability aspects in business strategies and operations. At this point, major business lines can be drivers; the sustainability department is not the major executer anymore.

9. Address new key entities
At some point, a systematic analysis of interfaces to sustainability and the next level of integration is necessary. A clear picture of rolling out sustainability to new organizational entities and employees has to be developed.

It is not easy to incorporate sustainability and make it truly a part of the structures and processes. Having key departments, such as research & development in a technology company or marketing in a brand company, or major business lines and anchor products aligned with sustainability, is a long way to go, but usually proves as the acid test. It might require proper incentives, deeper learning, relevant structures or a new mindset. For example, when key entities cooperate with critical stakeholders or experiment with startups in groundbreaking innovation, cultural change, as well as leadership focusing on new paradigms, is required.

To integrate sustainability at a deeper level, a heat map analysis can be used to depict the level of integration and necessary actions based on interviews with internal and likely external stakeholders.

10. Overcome organizational barriers

Understanding sustainability as a major driver in a change process will eventually lead to the point of facing what barriers are in place. The heat map will reveal challenges that counteract progress and full-scale integration. Often, external barriers are mentioned, as the CEO survey revealed (see pages 136f.). Yet, I like to argue that significant barriers are usually in place within the company. Said in another way: The internal aspects are the ones a company can influence.

The barriers will be as individual as the company. They could include a low level of motivation, no measurable goals, high ambition but a lack of implementation, too little intrapreneurial spirit, the absence of an executive role model, a lack of ownership, competing priorities between sustainability and business in the day-to-day routine and no adequate incentives. Organizational change measures will have to be discussed and implemented to overcome the major barriers.

ADDRESSING ORGANIZATIONAL BARRIERS

Some organizational and motivational aspects were already mentioned. Let me conclude this management chapter with a few final observations on change.

STATUS QUO—ON THE ROAD

Over the past 15 to 20 years, many companies, from big to small, have taken on sustainability. Where are advanced companies standing today?

– More structure: Dedicated sustainability teams and ambassadors are put in place at sites and in major functions.

– A broader reach: More topics relevant to stakeholders are addressed, underpinned by projects and linked to the business.

– A deeper scope: More suppliers are integrated and business solutions are created for major sustainability challenges, mainly risks.

– Increased focus: Strategies are developed and some targets set.

– Stronger accountability: Targets lead to measuring progress; ratings create transparency and comparability.

– Routinization of processes: Reporting and auditing become routine.

– Tighter communication: Reporting and communication on sustainability becomes increasingly high-quality, professional and insightful—yet also often too broad. Greenwashing is common (see chapter 5).

– Growth in sphere of interaction: Companies interact with more stake-holders, get involved and take their role in society more seriously.

– Permeation to the core: Sustainability starts to seep into the company's core, into its main products and starts to prompt a revision in business models, often pushed by industry dynamics.

– Recognition of cultural change: Sustainability slowly interferes with the company's culture, with developments in purpose and "new work".

This is where many advanced multinational companies stand today. They have come quite a long way on their sustainability journey.

A SOBERING DIVIDE

This chapter started out with the results of a CEO (Chief Executive Officer) survey. This survey interviewed more than 1,000 top executives across 21 industries and 99 countries about sustainability. This assessment is conducted every three years by the United Nations Global Compact and Accenture Strategy, with the most recent study dating from 2019.

According to the survey, leading executives of major companies in the world have recognized the importance of sustainability and have called for a decade to deliver. They recognize that ambitions are probably not yet sufficient and that business execution is not measuring up to either the size of the challenges or their previous level of ambition. Let´s cite a few specific results:

– 94% of CEOs say sustainability issues are important to the future success of their business.

– 76% say citizen trust will be critical to business competitiveness in their industry in the next five years.

– 78% believe we need to decouple economic growth from the use of natural resources and environmental degradation.

– 88% believe global economic systems need to refocus on equitable growth.

– Just 26% of CEOs in 2019 cited "no clear link to business value" as a barrier to sustainability compared to 31% in 2016 and 37% in 2013.

Yet, implementation is lacking, and business priorities oppose to sustainability.

– 55% operate under extreme cost pressure, which competes with investing in longer-term strategic objectives that are at the heart of sustainability.

– 43% have competing strategic priorities.

– For 63%, political uncertainty across markets is the most critical global issue for their companies' competitive strategies.

– Only 21% feel business is currently playing a critical role in contributing to the Global Goals, i.e., the SDGs.

– 28% see the absence of market pull as a major barrier and mention the lack of global regulatory frameworks to overcome them restraining progress.

The study concludes: *"The CEOs said in the interviews that their industries and business as a whole are not doing enough. Those same leaders agree that for the majority of businesses, awareness and commitment is not driving the level of urgency and concrete action required. They recognize—according to the numbers—that even their own businesses are not doing enough. While indicating a desire and willingness to do more, leaders say market constraints and an ever more challenging business environment and set of pressures continue to slow broad-scale transition to sustainable business—and that unless broader business is forced by a shift in economic incentives, action will stall."*[24]

HOW CAN THIS BE OVERCOME?

Without intending to overgeneralize or to simplify, a few points catch my attention when matching the CEO survey data with my long-year experience in sustainability. Let me suggest three organizational challenges to be addressed.

Challenge 1: Raise the ambition level to deliver.

Regarding ambition and delivery, my experience reflects precisely the results of the survey. The executives called for a "decade to deliver" but claim a low level of ambition and concrete commitments. They want to be active, theoretically. For the vast majority of businesses, awareness and commitment are not yet translating into the execution required for 2030.

Why? Many executives not yet fully believe that higher ambitions are necessary and resist committing to targets. For others, there is a fear of setting ambitious targets—either because they cannot be achieved under the given framework or they will not be rewarded.

Inertia and apprehension can only be overcome by facing it and competing for ambitious targets, likely by challenging each other and setting high bars. Executives should become a lot bolder, get more ambitious and dare to truly commit to accountable change. A good direction is given by the aforementioned commitment "The Climate Pledge" (see page 48) which states: *"Impossible is an attitude."* And *"It might sound ambitious—but that's the point."*[83]

Challenge 2: Get proactive on the changes clearly ahead.

The data delivers an interesting gap. On the one hand, the figures read: Change is coming and executives even agree: We need change. But the objections read: *"Our current strategies and business logic aren't suggesting that we do so."* Which is actually true, and matches the economic narrative drawn earlier (see pages 58ff.).

In many areas, market pull—due to consumer demands, regulations, rising costs, reflective pricing and pressure from powerful stakeholders—is lacking. Business pressures and market constraints have been awarded more importance by leaders. Sustainability issues, many of which are vague, feel far away and are not directly affecting business, lack priority in day-to-day business.

TIP

An interesting article challenging the status quo also from an inside perspective.
tinyurl.com/y2d44ozn

There are basically two approaches to solving this: either the pressure increases over time or businesses increasingly search for new ways to act regardless. The first would put companies eventually in a reactive position. The second would ask them to be proactive. Reactivity allows less freedom. Proactivity demands more courage. To put it simply, from a long-term perspective: You choose.

Challenge 3: Embark on change with a new way of leadership.

Business as usual doesn't lay the necessary groundwork for change. And sustainability isn't a new management fad. A profound change in management needs to take place that transforms business. This requires a new leadership style, which is also one conclusion of the CEO Survey. It addressed topics such as non-competitive collaboration, authenticity and making sustainability personal—some aspects I drew an own perspective on throughout the book. At this point, I would like to contribute to the emerging debate on transformative leadership a few provocative suggestions to prevent a cosmetic, superficial change.

 **TOOLBOX**

FIVE INGREDIENTS FOR A NEW LEADERSHIP STYLE

1. Stop underestimating sustainability; it is a complex challenge and highly relevant for business.

2. Stop overestimating your progress and sustainability performance.

3. Dare to question your economic narrative.

4. Break up silos within the company, industry and value chain.

5. Engage at eye-level with other people, including critics, underdogs, people affected by your business and passionate sustainability advocates from different contexts.

APPLICATION

Reflect on the following questions and write down what thoughts come up. Take your current corporate situation or a setting you are familiar with, for example, a prior job.

1. ——— What would be a trigger to get started with sustainability/or to take the next step?
– For you personally in your company?
– For your company as a whole?

2. ——— What do you know of that's already being done? How do you evaluate it? What is going well, and where is there room for improvement?

3. ——— Sustainability is often counterintuitive to what we have learned about business and how we do things. Where does this show in the context of your business?

4. ———— What barriers exist, and which incentives would work best in your setting? What would need to be done to get executives and key staff (e.g., in product development or marketing) to focus more on sustainability?

5. ———— Which companies do you admire for the way they integrate sustainability into their business? What do they do differently?

6. ———— What change would you like to see in your company?

7. ———— What can you do to make it happen, or at least to come closer to it?

AVOIDING GREEN-WASHING

When I talk about my business, I get often two reactions: admiration for a having a meaningful job followed, almost immediately, by the question as to whether the companies really care or are they just greenwashing. This default expectation of greenwashing poses a challenge for advancing sustainability.

Over the years, I have often seen communication that was indeed exaggerated, factually incorrect, and too self-confident: With a skilled view, I can easily detect mismatches between communication and performance. We see only the tip of the greenwashing iceberg. However, it is seldom a dumb, intentional cheat, as many assume. Often, it is caused by underestimating complexity and expectations due to little experience in the specifics of sustainability in companies and communication agencies. Given the fundamental lack of trust in the public, we need more bolder, humbler and sincere communication. The fine line is not easy to meet. We all benefit from learning to do so.

NEW TO GREEN-WASHING?

This whole chapter is about communications and sustainability. When combining these aspects, a lot can be done wrong. There is a fine line between recognizing the public's concerns and expectations correctly and addressing these appropriately. Let's first take a look at greenwashing—what it is, why it matters, and how it shows.

WHY IT'S NOT RANDOM

Greenwashing is a word with a negative connotation and judges a company's communication critically. Its roots are in the term "whitewashing", extrapolating it to environmental (green) topics. It usually refers to misleading communication: people feel tricked when a company tries to garner itself a green image and a good reputation. The typical approach is to over-emphasize certain positive facts or conceal facts that might imply a more balanced or critical view. Or it creates images, associations or an atmosphere that don't hold up to the facts. This is perceived as dishonest.

While it is often applied to green attributes, it can also refer to social aspects such as health, humanitarian care, employer qualities and other people issues. The mechanics would be the same as described above.

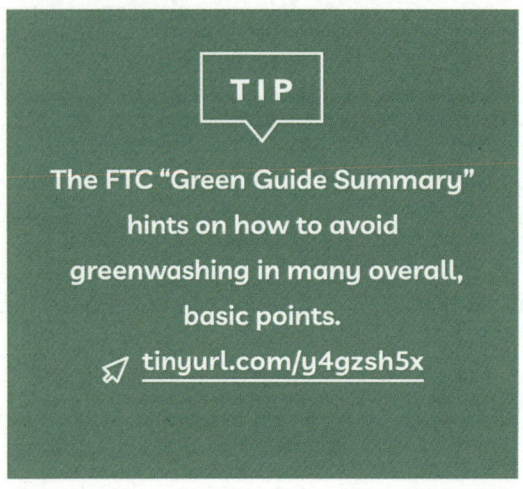

TIP

The FTC "Green Guide Summary" hints on how to avoid greenwashing in many overall, basic points.
⤴ tinyurl.com/y4gzsh5x

The results of greenwashing—if detected—are deception, loss of trust and, consequently, a lack of credibility. As trust and credibility are the most precious currencies in sustainability, greenwashing is a sincere challenge. Rebuilding trust after it has deteriorated takes a number of years and bold actions.

A guideline for marketers from the US-based FTC (Federal Trade

Commission) mentions some very obvious points about greenwashing. For example, if you use a label but do not meet the label's requirements or if you display a claim on your product that is a lie, this is obviously wrong. Most people would agree that this should not be done. That's a baseline.

In practice, it is often about fine lines and subtle aspects. If you search for the term greenwashing, you will find numerous websites with examples, some of them bashing on companies or presuming no good intentions. There are no objective criteria for greenwashing. Company's activities and communications are viewed from an individual perspective and measured, for example, by pre-knowledge, expectations, the opinion of the company, or a certain topic. What one person finds critical, another person might not. For this delicate topic, I would like to provide you with some, hopefully, non-judgemental guidance, being fully aware that it will be influenced by my own, personal perspective.

Let's start by looking in detail at two different examples showing the typical mechanisms of greenwashing and the pitfalls of reacting to it.

> **TIP**
>
> Although from 2010 and US-focused, the article shows an interesting range of "green-washed products" from different industries and describes features.
> ⊿ tinyurl.com/y9thmjs

WHEN OVERSHOOTING FIRES BACK

In 2007, the oil and gas company Shell published an ad in newspapers in Britain, the Netherlands, Belgium and Germany with the headline *"Don't Throw Anything Away. There is No Away."* It showed the outline of an oil refinery that had chimneys growing flowers. Part of the ad read: *"We use our CO_2 waste to grow flowers and our sulfur waste to make super-strong concrete. Real energy solutions for the real world."*[84]

NGOs, mainly environmental organizations, filed complaints against the ad saying that the environmental claims were likely to mislead readers; they also criticized the general depiction. In a complex process, regulators in different countries agreed on some aspects and dismissed others. In the

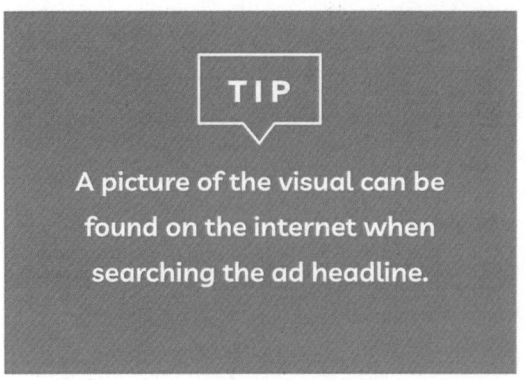

TIP

A picture of the visual can be found on the internet when searching the ad headline.

UK, Shell was required to withdraw the ad and eventually removed it from all markets. What happened?

Shell probably had good intentions with its projects to reduce carbon emissions: In one project, Shell supplied CO_2 waste to greenhouse growers who had previously burned natural gas to produce CO_2 to improve growing conditions. Also, it had developed a new method of using sulfur waste produced during the refining process instead of cement as a binder in concrete production. The problem was not the projects but the communication. Shell accepted the regulator's decision but found it disappointing, responding in a statement *"The advertisement was one part of a series that aim at opening and deepening the debate about meeting the energy challenge. We believe that the advertisement is a creative and striking way of drawing attention to the problem of waste disposal."*[85]

That's a fair and open reaction, but it reveals a lack of understanding about why the reaction was so critical. What can we learn from the case still today?

One major issue was commensurability. The flowers emerging from the chimneys referred to 17,000 tons of additional CO_2 being emitted that could be saved by the project. As for sulfur-strengthened concrete, 8 tons had been produced so far, with production expected to grow considerably in the next 10 years. Both were very small amounts compared to Shell's carbon impacts. The problem was: This was not explicitly revealed, and the overall wording suggested something different. That's where the ruling accused Shell of misleading the public:

"In the absence of qualification, most readers were likely to interpret the claim 'We use our waste CO_2 to grow flowers,' especially in the context of the image and the headline claim 'Don't throw anything away there is no away,' to mean that Shell used all, or at least the majority, of their waste CO_2 to grow flowers, whereas the actual amount was a very small proportion (0.325 percent of Shell's emissions), when compared to the global activities of Shell," the ruling

said. Regarding the waste sulfur claim, the authority said "readers were likely to interpret the claim 'We use ... our waste sulfur to make super-strong concrete' to mean that Shell used all, or at least the majority, of their waste sulfur to make super-strong concrete."[84]

The general communication approach suggests that Shell wanted to garner itself a green image. Newspaper ads are an epitome of widespread communications. Communicating a pilot project with comparably little impact via newspaper ads can easily overshoot. On top, the missing solid factual base and lack of sufficient contextualization undermined credibility. The visuals themselves were not ruled on by regulators. For many people, chimneys emitting flowers might be viewed as fun or harmless or "that's advertising". Let me point out that there is a difference between general advertising and advertising on sensitive environmental or social topics. From a sustainability perspective, it's the wrong visual language. And I would like to be very clear about that point as, over the years, we have seen flowers, trees and the alike coming out of chimneys, cars and airplanes, as well as nuclear power plants in lovely landscapes called climate savers. Most of them were lucky and their cases were not made public as in the example of Shell. Still this is no help to the topic, the company or the overall approach to avoiding greenwashing communication.

In general, it was the wrong topic for a company in the public eye. This often happens in greenwashing. A company with high emissions, severe environmental impacts and little credibility in the matters addressed in the projects needs to be especially careful when mass-communicating about a topic like climate change and its contributions to reducing it. It's an invitation for criticism. It needs to win over the skeptics.

WHY BRANDS NEED TO WATCH OUT

The brand Capri Sun was introduced in 1969 and is available today in more than 100 countries. Its product is a juice concentrate drink that comes in many different flavors and in a special pouch with a straw. It has been a companion for both children and adults for decades with its convenient pouch long before the to-go hype. Capri Sun guarantees that it uses only natural flavors and no preservatives or artificial sweeteners.

In 2013, Capri Sun was awarded the "Golden Cream Puff" by Foodwatch, a European advocacy group that focuses on protecting consumer rights pertaining to food quality. Foodwatch provokes the industry by awarding a negative prize to food manufacturers for *"the most brazen advertising lie of the year"*, as they state. Consumers vote for it, and the NGO publicly confronts the winning company in an uncomfortable manner.

Foodwatch frequently criticizes the approach of food brands to address children in marketing, especially with high-margin sweets and snacks. They used Capri Sun as an example, basically accusing the brand of seductive marketing to children.

The basic accusation against Capri Sun referred to the extensive amount of sugar in its products. Containing 12% juice and 10% sugar, adding up per pouch to more than 20 grams of sugar, this is not the healthy, re-commended drink for children for which it is marketed, Foodwatch stated.

The main accusation was that Capri Sun targets its marketing explicitly at children. It advertises aggressively to drink more of the sugary drink— targeting children in TV ads and social media, by sponsoring sports events, holding their own adventure camps and distributing promotional educational material in schools. In this material featuring product and company logos, the drink was misleadingly labeled as healthy in addition to the recommendation to consume a lot of it.

The company behind the brand (SiSi) reacted to the nomination of the award with a press release that was generally fact-based. For example, they pointed to regulations that foresee only 6% fruit content in a juice concentrate—a level the brand exceeds. They mentioned

> **TIP**
>
> The first video shows the sarcastic reaction of Foodwatch to SiSi's press release. The second video shows the intent to hand over the negative award. It demonstrates how critical stakeholders are campaigning. Both videos are only available in German but give an impression through the visuals.
>
> ⌁ vimeo.com/67287443
> ⌁ tinyurl.com/y2s6kngp

some low-sugar products in the portfolio and argued that its marketing doesn't address children but parents. This prompted Foodwatch to produce another video which confronted Capri Sun and consumers in a sarcastic way with the reality of its marketing, which is indeed aimed at children.

The lesson from this case that we can still draw today is that greenwashing also applies to social topics such as health and responsible marketing. When topics have high public awareness and a direct personal impact—like health issues and sugar intake have—the topics can be considered critical. Marketing to children remains a challenging arena. Nowadays, protecting children is given a high value, and aggressive marketing is not socially accepted.

Several industries, among them the food industry with sweets and snacks, are under general suspicion, and communication needs to be considered carefully. For a sweets or snack company, it is not genuine and, in some cases socially unacceptable to engage in advertising to children in connection with sports and related events. Accordingly, corporate educational materials for schools (now prohibited in Germany) is a very delicate way of engagement and should be carefully designed. As a result of the Foodwatch campaign, SiSi immediately withdrew its educational material from schools.

This case also shows that pure fact-based communication is often not sufficient. You might have the facts right, but without social acceptance, the facts will not be properly received. Denial of critical issues is often counterproductive and to critical issues can be a dangerous strategy for companies.

Over the years, the brand has expanded its communications about sugar as a result of the campaign and growing consumer awareness. In some specific areas, however, it still finds itself vulnerable to attack. Capri Sun has launched several low-sugar, no-added sugar and 100% juice drink varieties over the years. One motivator was parents' concern about their children's sugar intake, as well as more competition in the children's drinks market. The company's broad range of products has remained the same.

In campaigns related to social topics, authenticity and credibility are crucial. For example, years later in the US, Capri Sun featured an ad addressing social exclusion and mobbing in schools, see: ⤳ youtu.be/zN_Yrt5u0To An excellent and relevant topic—but from my perspective a questionable fit. It would have been better to have had more contained product placement and follow up on the topic over a long period of time, thereby raising credibility and making the commitment truly authentic. This is one of the fine lines that distinguishes sustainability communication from conventional advertising.

Take it as a footnote, that the environmental aspect of the pouch was not under scrutiny in the case. The company has addressed this issue, however, by reducing the overall material for the product and the foil layer. SiSi called its pouch eco-friendly in the past. This is an attackable statement: It contains aluminum and the layers of the pouch need to be separated for recycling—with the consequence that most of the billions of used pouches are thrown away, discarded as litter, or go into garbage without recycling.

SiSi now communicates on the website in a more contained way. [86]

> **TIP**
>
> The 15-point plan against misleading labelling and advertising published by Foodwatch can work as a good guideline on things to consider in the longer run or in specific marketing campaigns.
> ⤳ tinyurl.com/y3drha5d

Some brand companies and advertising agencies might not yet recognize the real point in this case or in the additional comments because, in fact, it is their usual way of doing business and marketing. I believe that over the next decades, the way of doing marketing will change dramatically. It will increasingly need to include sustainability aspects along the mentioned fine lines, rebuild trust, and authentically address critical topics. With the danger of greenwashing and the general scrutiny of the public, change will not only be a matter of communications but also of management and business transformation, as we saw in the previous chapter. Still, communication plays a strong role in the transition. Not all—probable few—greenwash approaches get noticed or receive major public attention. But I would recommend to avoid being the one on the radar.

TOOLBOX

CHECKLIST: WHEN TO WATCH OUT FOR GREENWASHING

The following points will help you identify whether you are in danger of greenwashing.

1. Are you in an industry generally seen as critical or in an industry under public scrutiny? Critical are the following industries that are usually excluded from socially responsible investing: gambling, pornography, weapons, and military activities. Some socially responsible investors also avoid other businesses perceived to have negative social effects. Among these are alcohol, tobacco, fast food, contraception/abortifacients/abortion and fossil fuel production. Industries under public scrutiny include the oil and gas, energy, automotive, chemicals, textile, pharmaceuticals, aviation, toys and some food segments. Be aware that public awareness is very powerful.

2. Do you advertise about a potentially critical or voguish topics such as climate change, waste, biodiversity, combating hunger or human rights? Do you plan marketing involving children?

3. Do you plan mass communication, e.g., TV ad, newspaper ad, social media campaign?

4. Do you deal with a pilot project or a prototype that is topic of your ad?

5. Do you lack specific numbers or have unproven facts? Do you plan to keep your numbers or results confidential to make your statements look better? Do you lack verifiable evidence from independent parties?

6. Do you plan to use specific words that are often connected to greenwashing, such as eco-friendly, natural, biodegradable, climate-friendly? Do you plan to make bold statements about your performance? Do you plan to use specific images that are often connected to greenwashing, such as images of nature and pictures of children?

7. Have you or a competitor ever been the target of a public campaign by an advocacy group?

If you answered "Yes" to more than one of these questions, my recommendation is to have an expert check on the greenwashing potential of your plans. If the answer to more than three questions is "Yes", you definitely should alter your approach.

5.02 ——— HOMEWORK IN COMMUNICATIONS

Communications on sustainability can sometimes be difficult. A lack of public trust, high expectations from critical stakeholders, a range of areas for tripping-up will challenge companies. Often, a balance in communications has to be found even within the company. Let's first look at one basic requirement: transparency. We will examine this quality through different lenses and at different stages.

SHOWING YOU ARE ON BOARD

These days, companies, especially larger ones, are expected to disclose at least some fundamental information related to sustainability: What is the attitude towards sustainability and certain industry-specific topics? How does the company address them—which activities are planned or already underway? Some basic communication is necessary to some extent.

Needless to say, to use the previous analogy, you first have to build your house well before you paint and decorate it. This means that it is necessary to establish some essential analyses and activities. This was described in the previous chapter. From that point, communications can take off.

I would like to present five points for starting your communication on sustainability. They are a summary from numerous communications projects I led with companies, often after initial analysis and setup of sustainability.

 **TOOLBOX**

FIVE STEPS TO DEMONSTRATE YOUR COMMITMENT TO SUSTAINABILITY

1. Define your who, why, what, how, and when.

a. Who: A brief description of the company's business model and the social and environmental impacts across the value chain
b. Why: Motivation to act more sustainably; thereby avoid stock phrases, underscore your commitment with corporate values or authentic personal motivation
c. What: Specific actions to address sustainability challenges, e.g., strategic approach (if applicable), focus areas or first initiatives
d. How: What has already been done and the way forward, including goals and performance indicators, compatible labels or standards
e. When: The roll-out schedule, for example by 2025 or 2030

2. Write a short summary of maximum 3 sentences defining your who, why, what, how, and when. Consistently repeat it across your ongoing communications, for example, as a boilerplate.

3. Set up a "static" passage on your website that unfolds into different formats. Social media, website blogs etc., work best, for example, for short project updates and new initiatives. Save press releases for bigger announcements.

4. Align subsequently basic information in terms of (a) wording, (b) the main messages and (c) performance data across all different communication formats, such as website, social media, reports, press releases and alike.

5. Walk the fine line: Avoid buzz words and generic phrases. Don't exaggerate or be too bold in your statements. Be authentic and in line with your overall communication approach, but more conservative in sustainability.

Later, you will work your way through other layers. For example, by expanding communication about key projects. Analyzing risks in your product marketing and adjusting it to sustainability requirements. Involving stakeholders and publishing a materiality analysis after the stakeholder survey. Posting statements on crucial topics. And so on. Communications should be in line with the progress of your activities and must be crafted to the specific circumstances of your company and industry.

 **TOOLBOX**

WORDING TO AVOID

Consider the following examples for wording to avoid or use properly.

1. Buzzwords: sustainable, agile, integration, synergy.

2. Generic phrases: We continuously improve our sustainability performance across our business. People-Planet-Profit has the highest priority for us. Our way of doing sustainable business contributes in various ways to our economic success.

3. Exaggerations: We act responsibly everywhere. All of our employees are involved in sustainable businesses.

INTERACTING WITH STAKEHOLDERS

At a more advanced point on the sustainability journey, engagement with stakeholders has to start which has many facets. BMW for example uses multiple ways and formats to engage with different stakeholders— dialogue, conferences, discussions, student forums, surveys, plant tours, digital formats and so on. See: tinyurl.com/y48tr4jt

It demonstrates an advanced state as well as an intensive and different-iated interaction. For a start, choose two or three stakeholder groups and formats to eventually expand on for the purposes suited to your context.

Pharmaceutical company Novartis provides statements on key topics to increase transparency. As the company describes it:

"In order to find long-term, sustainable solutions to global health challenges, we need to engage in constructive discussions and be transparent about our views. We engage in policy debates and share the company's perspective on issues including access to healthcare, ethics, R&D and quality of medicines, striving for open and constructive interaction. As part of our ongoing commitment to transparency, we state our position on key issues affecting our business and of specific interest to our stakeholders."[87]

The website includes short papers on Novartis' positions on access to healthcare, animal research, collaborating with patient organizations, pharmaceuticals in the environment, ethical principles for transplantation studies, its tax policy statement and others. In doing this, the company demonstrates a high level of transparency, even about highly controversial topics that most companies still are hesitant or reluctant to address.

TIP

Learn more about Novartis Social Business.

tinyurl.com/yyfsmebt

An even broader approach may be appropriate for crucial topics. To address the topic of the access to healthcare—a critical topic in the industry—the company formed the special entity Novartis Social Business (NSB). NSB works in local communities with partners to provide affordable, high-quality medicines against infectious and chronic diseases while strengthening healthcare capacity in lower-income countries. Regarding the issue of how to improve access to healthcare in lower-income countries, NSB hosted several stakeholder dialogues that could be followed on Twitter #TalkingNCDs. ⟋ tinyurl.com/yylfu8lo

C&A is another example of a company practicing high transparency. For their supply chains, few textile companies report concrete numbers or approaches. C&A sets a standard for the industry with its transparency on its website and in its report. ⟋ tinyurl.com/y3txtn3y

Transparency is understood as a prerequisite to monitoring suppliers and improving conditions in the supply chain: *"Our overall approach: An important first step to achieving greater transparency in our supply chains is to make sure we are gathering accurate data about our suppliers' performance—for everything from chemical, energy, and water use, to issues in relation to wages or safety in the workplace—and assessing their ongoing actions and results. The more comprehensive and accurate the information is, the more targeted and effective support we can offer."*[88]

C&A discloses that the supply chain encompasses more than 1 million people, employed through 722 global suppliers who run more than 1,600 production units across four entirely different sourcing regions. The company has maintained relationships with over 71% of its suppliers for more than 5 years and, within the last two years (referring to 2018), has reduced its supply base by 39%. C&A discloses in detail how it assesses suppliers, its goals on improving the supply chain, conducted third-party audits and its progress in different countries. This not only allows them to manage the supply chain well and invest specifically in expanding capacity, but also helps to build credibility for the company's sustainability efforts. High transparency also raises the bar by exposing what is possible for a big textile company to and, in turn, challenges peers to follow suit in terms of transparency and performance.

SUSTAINABILITY IN SOCIAL MEDIA

Social media is the leading source of information for young people. How can this channel be used for sustainability? At first glance, the fit is not very obvious as social media has short posts and is generally visually focused. Sustainability, on the other hand, is usually complex and nuanced, requiring a lot of explanation, reports are long, and performance data needs a larger context.

A consultancy on sustainability called Context conducted a study in 2019 on how companies were currently using social media, specifically Twitter, for sustainability. The basic data and methodology can be found in the article.[89] I would like to summarize some of the major findings of this study below.

– Companies use a blend of social media platforms (Facebook, LinkedIn, Instagram), but Twitter is most commonly used by sustainability teams.

– On Twitter, either the main corporate handle, dedicated sustainability handle or the personal handle of the sustainability leader is used.

– The advantage of posting sustainability content using a main corporate handle is to demonstrate mainstreaming of sustainability within the company. As it usually has the most followers, a broad audience sees sustainability posts.

– Sustainability influencers, in comparison, are more inclined to follow dedicated sustainability handles or, to an even greater degree, personal sustainability leader handles. The top seven personal sustainability leader handles actually have more influencer followers than their companies' corporate handles.

– Posting frequency was highest on dedicated sustainability handles.

– According to the analysis, among the most interesting companies to check out for their social media performance in sustainability (with some focus on Twitter) are Cisco, Microsoft, Johnson & Johnson, Marks & Spencer, the Hewlett Packard Enterprise, Autodesk, The Home Depot, Apple, Nestlé, SAP and Siemens. Some of these are corporate accounts and some are personal accounts.

GOOD REPORTING ON SUSTAINABILITY

In the European Union, large companies are by law required to disclose certain information about the way they operate and manage social and environmental challenges. The "Directive 2014/95/EU" is also called the non-financial reporting directive as it requires the disclosure of non-financial aspects. Specifically, the regulation requires the disclosure of the management approaches and performance regarding the relevant aspects in environmental protection, social responsibility and employee-related topics, as well as on human rights, anti-corruption and bribery.
⌁ tinyurl.com/yahhcwng

Other reporting frameworks exist for sustainability reporting such as the guidelines for reporting of the Global Reporting Initiative (GRI) and the Integrated Reporting Council (IIRC) that combine financial and sustainability into so-called integrated reporting.

Additionally, non-financial reporting on ESG criteria (Environmental, Social, Governance, see page 55) is part of investments and is therefore expected by companies listed on the stock market in particular. Several rating companies assess and score performance. In most cases, companies are automatically rated based on disclosures in the public domain, regardless of whether the company desires such ratings or not. The ESG scores are then sold by the ratings firms to interested outlets. Through this system, the level of transparency is growing quickly. Indices such as Dow Jones Sustainability Index (DJSI) or the FTSE4Good Index series list companies according to their sustainability performance based on a range of corporate social responsibility or sustainability criteria.

Further initiatives participate in the goal to increase transparency. The measurement and reporting of greenhouse gas (GHG) emissions is becoming widespread across industries and countries as society attempts to reduce GHG emissions. The Carbon Disclosure Project (see page 86), for example, requires transparency, especially from companies in high-emitting industries such as the oil and gas industry. The industry itself already published its first sector guide for reporting in 2004, which has continued to be deepened and improved.

> **TIP**
>
> The Sustainability reporting guidance provides direction for the oil and gas industry by covering 21 sustainability issues and 43 indicator categories.
>
> ✒ tinyurl.com/yxlfyex5

In an effort to control sustainability in their supply chains, companies increasingly request information from other companies, mainly through customer questionnaires. Stakeholders demand transparency and monitor industry-specific topics. One example is the WWF which uses its Palm Oil Score Card to rank companies from the dairy and meat industries to prevent deforestation in palm oil supply chains (see page 96). Another example is the Access to Medicine Index, which publicly discloses the performance of companies in the pharmaceutical industry (see pages 217f.).

Reporting has become an extensive field. It consists of more than publishing a sustainability report, though this is often at the core. There are very different requirements connected to the variety of reporting requirements mentioned above. A challenging task for advanced companies is to streamline reporting while making it efficient and effective at the same time.

 **TOOLBOX**

FIVE MAJOR POINTS FOR GOOD REPORTING ON SUSTAINABILITY

1. Find your reporting arena

Define the scope that applies to your company based on the different reporting sources and guidelines ranging from GRI report, ESG reporting, investor scorings, CDP, the EU CSR directive, customer and supplier questionnaires, and management system requirements (e.g., ISO 14001, OHSAS, SMETA etc.) to particular industry-specific rankings and scorecards, such as the ones mentioned above.

2. Understand the requirements

Each of the sources mentioned has their own different requirements. Some customer surveys comprise over 100 questions, a GRI report requires a materiality analysis, the non-financial reporting directive calls for the definition of certain key performance indicators and due diligence processes, and industry rankings require certain scopes. These have to be analyzed and streamlined.

3. Prepare the data collection

Usually, many different departments must deliver data for the reporting universe, especially in multinational corporations, data collection requires some preparation regarding scopes, responsibilities, availability and processes—especially as data differs along the different requirements. For several requirements, qualitative statements have to be defined that hold up to external revision. This is a longer, iterative process.

4. Streamline data into different funnels

As most data is required on a frequent base, e.g., yearly, an automated setup for data retrieval is recommended. Existing technological solutions must be assessed for the likely integration of new data. Automated processes, for example, for answering customer

> questionnaires must be defined. Sending out consistent data to diffe-
> rent systems in an effective and efficient manner must be ensured.
>
> **5. Translate performance into further communication**
> The extensive data that is collected in the reporting requirement pro-
> cesses in the backend can be intelligently used for communication on
> the frontend. Defining a yearly communication plan, including which
> performance data is requested by the public and which is desirable
> by the company, helps to make the best use of the data available.
> Naturally, the whole process of data collection, analysis and com-
> munications has to be assessed regularly for accuracy, progress and
> necessary adjustment.

Chemical company BASF has been progressive in integrating sustainability into its management and reporting for many years and has participated, for example, in the CDP from the very beginning. The company's
current strategy is based on eight non-financial targets, including the
most material topic: BASF's commitment as an energy-intensive company
to energy efficiency and global climate protection, striving to reduce
emissions across the value chain. Reporting is very detailed and
worthwhile checking out for advanced companies.

⟋ tinyurl.com/yyprc968   ⟋ tinyurl.com/y2szsvnm

5.03 ——————— **MASTER CLASS FOR VOGUISH TOPICS**

I mentioned earlier that voguish topics potentially qualify for greenwashing. There is no clear definition of voguish topics. Their relevance
differs by industry, country and over time. Currently, climate change,
organic food, waste, e-mobility, human rights, diversity and anti-racism
are examples of likely voguish topics. This does not imply any judgement—
each of the topics is highly relevant! What is meant in relation to our topic
of greenwashing is that these topics entered the mainstream and have a

high level of attention and public awareness. Usually, collective expectations of and sensitivity to these topics are high. People will critically scrutinize company's activities in these areas as suspicion of greenwashing is high. So, it is suggested to pay careful attention to communications—especially advertising—surrounding voguish topics in sustainability.

Companies want to avoid negative associations. The goal would be to have solid, transparent communication that raises attention and improves the reputation of the company. Let's work through a specific example and the challenges related to communication.

THE CHALLENGE: PLASTIC POLLUTION

In recent years, several companies introduced ocean plastic, beach plastic, marine plastic and social plastic into products and packaging.

TIP

Watch the Coca-Cola video regarding its ocean plastic activities. Does it meet the fine line of sustainability communications or is it greenwashing for you? Why?
⏎ youtu.be/2sodwfv1g9E

After that you may watch an excellent critical documentary on plastic waste and Coca-Cola. The trailer is availabe for free.
⏎ tinyurl.com/ydxmp2us

– Adidas offers several ocean plastic sport shoes.

– Head & Shoulders (Procter & Gamble) introduced the world's first recyclable shampoo bottle made from beach plastic.

– Coca-Cola, the biggest plastic packaging producer of the world, piloted a small number of bottles made from 25% marine plastic.

– Henkel announced the integration of social plastic into packaging for beauty care and laundry home care products. It also uses the term "ocean plastic" for a number of products.

Four different types of wording and approaches for basically the same challenge: plastic pollution. While it is noble and necessary to address it, proper communication on this voguish topic is challenging, as practice shows.

PLASTIC POLLUTION

Half of all plastics ever manufactured have been made in the last 15 years. Production increased exponentially, from 2.3 million tons in 1950 to 448 million tons by 2015. Worldwide, 1 million plastic bottles are used—per minute. 91% of them are not recycled.[90]

Production is expected to double by 2050. Every year, about 8 million tons of plastic waste escapes into the oceans from coastal nations. Millions of animals are killed by plastics every year, from birds to fish to other marine organisms. Nearly 700 species, including endangered ones, are known to have been affected by plastics. Most of the animal deaths are caused by entanglement or starvation.[91]

Currently, there are five huge plastic islands that have formed throughout the world: in the North and South Pacific, in the North and South Atlantic and in the Indian Ocean. The following video explains how this happens and why it is a problem. vimeo.com/154149905

The Great Pacific Garbage Patch (GPGP) is one of them. See the video for pictures and explanations. tinyurl.com/y4j5cyzt

Plastic pollution—and with it, microplastics—is one of the world's major environmental challenges. You can find detailed data and some promising initiatives at the following links. Scroll also around Instagram for current discussions.

tinyurl.com/y32e33jd   tinyurl.com/y32e33jd

tinyurl.com/y4znx2ho   tinyurl.com/mc6aqpa

endplasticwaste.org

WHY WORDS MATTER

We saw earlier that the wording applied in mass communications is crucial. This is even more so for voguish topics. In the specific example, there is not yet a clear definition. Recycled plastic is a defined term. But ocean, beach, marine, and social plastic are terms created without a clear definition.

When presented with a shower gel that prominently markets the term "ocean plastic" on the packaging, consumers in Germany in a survey assumed concordantly that plastic was taken out of the oceans, recycled and used to produce this bottle of shower gel. This is not surprising considering that the wording "ocean plastic" suggests exactly that. People are aware of oceans full of plastic waste and sea birds dying. These pictures and messages are now anchored in the minds of a broader public because they have been communicated widely—a typical characteristic of a voguish topic.

"Ocean plastic" relates—probably intentionally—to this association. Adidas also consistently uses the term ocean plastic and applies visuals supporting exactly this association (oceans, floating plastic garbage, marine wildlife). youtu.be/9lui-pGjtZQ

However, the wording and visual language in both cases is wrong and plays on people's associations. The plastic used is not taken from the ocean and does not contribute directly to reducing maritime debris. The respective plastic waste typically comes from beaches and coastal regions and is collected before it enters the ocean for recycling. This is what companies explain in more detail, for example, on their websites. From the strict perspective of avoiding greenwashing, the wording and visual contextualization, however, are not accurate.

Consumers taking the survey in Germany became very angry and felt betrayed when they heard the genuine facts. The reality is that recycling and manufacturing plastic waste from the sea is (still?) outrageously expansive and challenging in terms of material requirements and composition of the ocean plastic waste.

This is why some companies use the term "beach plastic" because the plastic is not taken from the ocean but collected from beaches and coastal areas. It is actually a more accurate term but probably less intuitive. However, it can be well contextualized, as waste on beaches and in coastal areas can be easily visualized and explained.

"Social plastic" is a still less intuitive term but also factually more accurate than ocean plastic. The focus here is on the plastic collected by people in coastal regions—either by volunteers to raise awareness of marine pollution or by garbage collectors to make a living. For a company to build an authentic story around social plastic, the recommendation would be to install specific long-term projects in affected countries.

The marine plastic Coca-Cola used in its pilot project refers to plastic taken out of the sea and is, thus, worded appropriately. But generally, the criteria indicating potential greenwashing (see page 151) were met by the company—which is why this communication approach, in my view, is overall not appropriate.

THE TRUTH AND NOTHING BUT THE TRUTH

Companies should communicate frankly about their amounts and approaches. Don't leave details out but state what you are doing clearly and correctly. Critical stakeholders complain that products are not sourced 100% from recycled plastic waste but typically mixed with conventional plastic.

Head & Shoulders' product packaging read: *"Meet the world's first recyclable shampoo made of beach plastic"*. ✈ tinyurl.com/y5ns27e9

It turned out that 25% of the material is recycled beach plastic, as it states in the detailed explanations: *"Plastic coming from the marine environment is notoriously difficult to re-use due to degradation, but working with its partners, Head & Shoulders has found a technically revolutionary way of integrating 25% of this plastic into its packaging. It is hoped the launch will not only help to clean up plastic on beaches but also inspire consumers to play their part and recycle their shampoo bottles and prevent more waste landing on beaches."* [92]

By curtailing information, the right facts did not reach the mass audience, which can feel misleading. In this specific case, it would have been fairer to state on the bottle either "... integrating beach plastic" or accurately state "Made from 25% beach plastic". Some people might find that nit-picking. Not at all when it comes to voguish topics or general issues in sustainability. Trust is fragile and needs to be earned. For the sake of companies and customer acceptance, clear and transparent communication is the best policy.

Another delicate point that needs to be communicated with more transparency is pricing. Recycled polyester costs about 10% more than the virgin material, according to Adidas. They argue: *"Adidas ultimately wants to get the price down so more consumers can afford to choose sustainable products."* Considering the overall cost structure of a sport shoe and a price of 150-200 euros for its ocean plastic sport shoe, 10% extra in polyester costs does not add up to a large amount. It sounds like overstating an argument in order to appear noble and responsible. No need for that from such an ambitious brand.

FORTIFY WITH GOOD GOALS

Waste reduction and packaging are material topics for consumer goods companies. Many companies give context to their application of recycled plastic in this area using overall goals, which is a good approach.

– Procter & Gamble, the parent company of Head & Shoulders, announced in 2018 its ambitious plan to introduce 25% recycled plastic across 500 million bottles sold yearly by its hair care brands.

– Henkel has set itself the goal of reducing the amount of virgin plastics from fossil sources in consumer products by 50% and increasing the share of recycled plastics by more than 30%. By 2025, all of Henkel's packaging will be recyclable or reusable.[93]

– Coca-Cola has made a global commitment to make all bottles with at least by 50% recycled plastic by 2030. By 2025, the goal is to ensure that all packaging is recyclable.[23]

– Adidas made more than 11 million pairs of shoes using recycled plastic in 2019 and plans to make 15-20 million in 2020. However, this is still only a fraction of the group's total of more than 400 million. Adidas says it wants more than half of the polyester it uses to be recycled in 2020, ramping up to 100% by 2024. In 2019, Adidas expected 46% of the polyester used in its clothing to be recycled, compared to just 28% for its shoes. [94]

Setting ambitious goals helps to align and focus corporate efforts. By committing to a journey, a company can be made accountable and refute complaints such as using recycled plastic for only a fraction of products. However, also here, sincerity and a clear plan forward needs to be presented. In some cases, double standards are assumed by critical stakeholders, as the video above questions for the case of Coca-Cola. (see page 162)

FINALLY, TYING IT ALL TOGETHER

For the moment, there is often a mismatch between advertising and other communication. It is the same challenge seen in traditional branding— emotions are played on, facts are curtailed and not taken too seriously.

In the case of Adidas, the visual appearance of the advertising suggests that oceans are cleaned up (see page 164)—which is not true. However, in its overall communications, it explains accurately how it collects and uses plastic in its products. This should set the course for advertising, too.

"Parley Ocean Plastic® is created from upcycled marine plastic waste that is intercepted on remote islands and beaches and in coastal communities. We use it as a replacement for virgin plastic in all adidas x Parley high-performance sportswear. After plastic trash is collected from coastlines, it's baled and sent to Parley supply chain partners. There it's shredded and reworked to become high-performance polyester yarn: Parley Ocean Plastic™, which is then used to create adidas x Parley sportswear that's as good for the planet as it is for your workout." [95]

Henkel did a good job from a sustainability standpoint with branding and the press release. *"The bottles of these Beauty Care and Laundry & Home Care products are made of 100% recycled plastic—of which up to 50% is Social Plastic."* [96] Though criticized for their use of ocean plastic in packaging in the German documentary mentioned above, they did several things right: They used correct wording (social plastic). They combined recycled plastic with social plastic and explained the break down. In their press release, they described in detail their approach to "integrating it into packaging", in which products they use it, a timeline, their partners, and their goals.

To sum it up: Communication connected to sustainability should have especially high standards for facts and accuracy—as well as for visuals. A well-measured contextualization and an alignment across all ways of communication, especially aligned with branding, is necessary. It will be exciting to see how new ways of inspirational, fact-based advertising may surge.

5.04 ———————— HOW TO NAVIGATE IN DIFFICULT TRANSITIONS

Often, the conditions for communications are not ideal. Companies may operate, for example, in industries perceived as critical and unsustainable. Or they may be struggling on their path to becoming more sustainable. They could lack credibility based on past behavior. There is often a bias within the company that hinders progress as not all people and activities are well aligned.

I would like to look with you at three examples of where companies have embarked on a path to creating trust-building communication under very difficult circumstances. The first two involve companies that are avoided by strict socially responsible investors as their businesses have negative social effects. All three are not doing everything right. Their business models remain controversial. Some readers might be challenged reading through them as good examples.

From my worldview, it is necessary and valuable to remain open to people´s and company´s efforts to change. And I believe that we can learn from companies in difficult circumstances as these companies require more courage and perseverance to embark the sustainability road. Let´s identify for each case a few components of long-term oriented sustainability communication under pressure.

CAN A TOBACCO COMPANY BECOME SUSTAINABLE?

Philip Morris International (PMI) is a multinational cigarette and tobacco manufacturing company with products sold in over 180 countries.

There are some 1.3 billion tobacco users globally, according to the World Health Organization (WHO). With tobacco being addictive and the single greatest cause of preventable death globally, according to the WHO[76], the core business of the company is highly controversial. Additionally, it has a history of obstructing scientific evidence surrounding the health impacts of smoking and is the subject of numerous litigations.

Could such a company ever become more sustainable? Would people place trust in their willingness to change? I would like to describe three approaches of Philip Morris.

First, the company today has a generally open, proactive communication style. Controversial questions and topics, such as the health effects of smoking or the "Ten questions skeptics often ask PMI", are not ignored. At the request of a stakeholder, PMI assessed its marketing approach and published a publicly available report on responsible marketing practices with an invitation to comment and challenge the company. It released videos with testimonials of people who had quit smoking. ⊿ pmi.com/smoke-free-life

> **TIP**
>
> An investigation on PMI's social media campaigning on IQOS, which the company stopped later.
>
> ⊿ tinyurl.com/y533t844

Secondly, the company has a clearly defined and communicated vision. They know that transformation is key to keeping its license to operate. "A company in transformation towards a smoke-free future"—this is a bold and unusual vision for a cigarette manufacturing company. It is extensively explained and supported. Philip Morris has been on this path for many years now and has become much more firm, clear and frank over time.

Transformation takes time, and trust is built through concrete evidence. PMI continues to rebuild the company based on the expectation that cigarette sales in many countries will cease within 10 to 15 years. The switch to alternative products to nicotine is a major approach and the third to mention. The company has set a target of having at least 40 million of the adult smokers who have stopped smoking, switch to its smoke-free products by 2025. Philip Morris estimates nearly 14 million consumers were IQOS users at the end of 2019, with almost 10 million having quit cigarettes. Also in terms of its communication, Phillip Morris has gradually developed into a more open and cooperative actor.

⚐ pmi.com/our-transformation

AN ALCOHOL COMPANY PROMOTING MODERATE USE

Diageo is a multinational alcoholic beverage company operating in more than 180 countries. Its brands comprise Smirnoff, Johnnie Walker, Baileys, Gordon´s Gin, Captain Morgan, Guinness and many others. Its aim is to be one of the best performing, most trusted and respected consumer products companies in the world.

The harmful use of alcohol contributes to 3 million deaths each year globally, as well as to the disability and poor health of millions of people, according to the WHO. A total of 1.2 million lives worldwide are lost every year due to drunk driving, which is 100% preventable.[97]

Diageo has set three priorities within the scope of their sustainability strategy: promoting positive drinking, building thriving communities and reducing environmental impact. Let´s focus on the first one, as it is the one that affects our societies the most. Here I would like to describe three of Diageo's approaches.

This first involves Diageo's commitment, together with other leading global producers of beer, wine and spirits, to the global goal of the World Health Organization of reducing harmful alcohol usage by 10% worldwide by 2025. This is a specific goal that the company can be held accountable for.

A second approach involves information, education and encouraging responsible choices by consumers. The baseline is the responsible marketing and digital codes on products. All labels and packaging in all geographies (where legally permitted) and all Diageo-owned brands have information on at least one, and up to three, of the following responsible drinking symbols: Do not drink and drive//Do not drink during pregnancy//Do not drink under the legal purchase age. The link to the company's website promoting responsible drinking DRINKiQ.com is on the packaging.

⭧ www.drinkiq.com

A third approach tackles the misuse of alcohol and consists of programs and training for ambassadors of responsible drinking, such as Decisions VirtualReality and #JoinThePact, as well as Smashed to combat drunk driving and underage drinking. The global #JoinThePact program aims to encourage 50 million people by 2025 to never drink and drive by signing a global pact. Ambassadors promote moderation through DRINKiQ and through brands such as Haig Club, Captain Morgan and Crown Royal.

For several projects, progress is measured. For example, the flagship theater-based education program "Smashed" informs young people about the dangers of underage drinking. Over the past ten years, "Smashed" has reached 700,000 young people on six continents. 95% of students knew more about the dangers of underage drinking after experiencing "Smashed." Diageo has a project portfolio across the world communicated via website and YouTube videos. ⭧ tinyurl.com/y979ozba

DARING A RE-START

British Petroleum (BP) is a multinational oil and gas company and among the world's largest in the industry. Through their extensive impacts on environment and societies and significant economic influence, oil and gas companies have been under strong scrutiny for many years.

BP was an early sustainability mover in the industry after extensively focusing on renewable energies, among other things. The company re-branded itself in 2001 to "Beyond Petroleum", emphasizing its ambitions to focus on the renewable energy trend and become an overall more environmentally friendly company. I had the chance to work with the German branch of BP in 2005 and experienced them back then as very professional and sincere about a new role as corporate citizen. Due to different business developments over the years, in 2011 BP withdrew from the renewable business and from its transformation path overall. A devastating major oil spill in 2010 in the Gulf of Mexico severely damaged BP´s image. As a consequence, the company focused its investment entirely in the traditional oil and gas business.

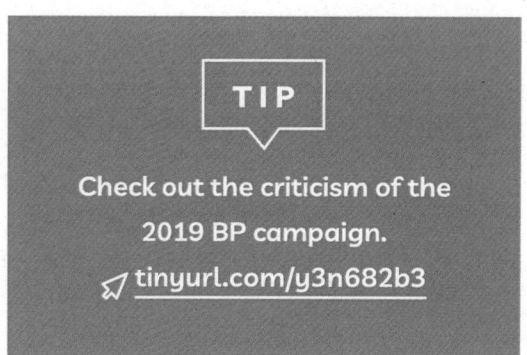

TIP

Check out the criticism of the 2019 BP campaign.

tinyurl.com/y3n682b3

In 2019, a new campaign was launched that tried to revitalize BP's image as a company seeking to advance, including in renewables. The global advertising campaign was called "we see possibilities everywhere". It aimed to showcase BP's efforts to embrace clean energy and included a series of short videos profiling BP's plan to increase its energy production while lowering its emissions. The campaign using major PR agencies was rightly perceived as greenwashing by a critical watchdog.

Aside from the only somewhat successful history and the "not well done" advertising campaign, I want to focus on a recent and promising approach of BP.

On February 12, 2020, in the second week after Bernard Looney took the helm as new CEO, the company set a new goal: to become a net zero company by 2050 or sooner, and to help the world get to net zero. *Reimagining energy—for people and our planet"* was stated as the new purpose of BP.

Mere public relations? Greenwashing?

It truly is a bold statement for an oil and gas company at a time when many bold statements are made. It may sound questionable based on the brief background description. There are three aspects that (could) make a difference.

First, there is top level commitment. CEO Bernard Looney has worked for BP for decades and loves his company. Still, it has to change, he states, and that there is no turning back. His inaugural speech on the new direction is truly worthwhile to listen to. It provides hope that the new direction and the transformation of BP can be implemented. ✒ tinyurl.com/y2esezzq

Secondly, BP sets clear and ambitious goals. Under the goal of becoming a net zero company by 2050 or sooner, BP has defined 10 goals—five for transforming the business internally and five for contributing to societal transformation. Accomplishing these will mean tackling around 415 million tons of emissions—55 million tons from BP´s operations and 360 million tons from the carbon content of upstream oil and gas production. It is striving for absolute reductions, to net zero. It aims to cut the carbon intensity of the products it sells by 50% by 2050 or sooner. The amount roughly equals the total emissions of the UK. BP includes the scope 3 emissions in the value chain, which is ambitious.

Thirdly, BP has learned from past hard experiences. The company and its CEO are aware that they will be held accountable. They know their weaknesses and about critical stakeholders' lack of trust. That was made very clear by the CEO when announcing the new vision in his inaugural speech. They stopped the aforementioned controversial advertising campaign "we see possibilities everywhere". BP committed to ending lobbying against their set purpose, which is novel, and stated that they would be open for any comment on not holding up to what they promised.

Obviously, all this is not a guarantee. And leading a business like BP's into transformation during times of turmoil is highly demanding. But from all the parameters, especially associated to personal accountability, it sounds promising. And I am convinced that when someone fails and then dares to start anew, they have the power to succeed because they know what's at stake. BP started the new transformational path in an ambitious, professional and open way. May their courage and honesty succeed, multiply and be rewarded.

APPLICATION

1. ———— Flip through the examples in the links. Which examples do you personally find critical and would judge as greenwashing? Which occur less problematic to you? What does this reveal about your preferences?
   ◁ tinyurl.com/y6rusjnx
   ◁ tinyurl.com/yy5uqc68

   How does this reflection alter your thinking about the term of "greenwashing"?

2. ———— Think of a company/product that won you over with its sustainability communication. What did it take to gain your trust?

3. ———— Are there companies you generally reject and would not trust, no matter what they did? What would have to happen for them to gain your trust?

4. ———— As a corporate communicator, how would you ensure that a customer would assume your communication is trustworthy?

– An automotive company wants to promote its new e-mobility flagship model. The vast majority of brands has basically no sustainability features. The company has a good track record in sustainability with regard to its own production. How to avoid greenwashing in the promotion of the e-mobility flagship?

– A cosmetic company uses mostly plastic as packaging and is attacked by critical posts on social media. Switching to natural packaging solutions is currently out of scope—it is either non-existent or too expensive.

– A dairy company launches plant-based alternatives. How should sustainability concerns about their core business—milk—be addressed in their overall communication while promoting TV and social media ads on their new plant-based products?

5. ———— How could you as a consumer and responsible citizen
– request more transparent information on a certain topic/industry?
– give companies more encouragement to advance their sustainability agendas?
– share good examples in your circle of influence?

IDEAS AND INNOVATION

I have worked for several years with numerous students on ideas for sustainable businesses. Each year, the task was to develop business plans for challenges in our world. Students had the option of coming up with something from scratch as startup entrepreneurs, or as intrapreneurs developing sustainable innovations for established big corporations. Many great ideas for real problems emerged from that exercise.

What was so fascinating about this lab atmosphere for me as well as the students? It was the unfolding of creativity and energy—when ideas developed into solutions. Joy and passion for entrepreneurship flourished. Young people became knowledgeable, self-confident and proud as they presented their final business plans. The aspiration to exert influence through business and to make a difference were encouraged. It was like watching a wonderful plant grow from a seed in fast motion.

Ideas drive business, move our world and fascinate people. Let's start at the beginning and look at the impact ideas can have and try to find inspiration from a broad range of ideas involving the multiple facets of sustainability.

CREATING TO IMPACT

The passion to tinker, the impetus to change things and the creativity released in innovation processes are among the most fascinating qualities and distinct characteristics of human life. Dreaming and creating are inherent to humans.

Some ideas have changed the course of our world—think of inventions like the letterpress, electricity, telegraphing, railways, airplanes, penicillin, and the internet. Their impact on the economy, culture and social interaction was tremendous. Inventions that make life flourish more are especially powerful. Still, inventions often develop negative side effects over time that need to be managed—which is another facet of the creation processes.

Sometimes, ideas "only" change the world for particular individuals. So does VerbaVoice, a German company that started up in 2009. VerbaVoice makes it possible for speech-to-text reporters and sign language interpreters to work location-independent. Now these professionals are affordably available without requiring the physical presence of the interpreter. For people with hearing disabilities, this, in fact, makes the world accessible.
⟋ www.verbavoice.de/english

TIP

The Ocean Cleanup project publishes interesting posts on progress.
⟋ tinyurl.com/y4osr4ht

An article summarizes scientist's critics to the project.
⟋ tinyurl.com/y6xmqb3k

Sometimes, monumental problems are met with bold ideas, such as "Ocean Cleanup". It sounds a bit like a modern fairy tale: An 18-year-old who aims to clean up plastic pollution in the ocean using technology forms a nonprofit and raises money for his prototype through a crowdfunding campaign and starts piloting the idea. It is still unclear whether the project Ocean Cleanup will succeed from the standpoint of both economic viability and environmental effectiveness. Nevertheless, the founder and his team still try. The immense challenge of plastic pollution has already drawn broader awareness, prompting other initiatives such as Plastic Bank to join the playing field.

Technology facilitated another idea that also sounds like something out of a fairytale: A 60-square-meter house was built in one day and cost only 4,000 US dollars because it came from a 3D printer. The first such houses were built in 2019 in Mexico and El Salvador by New Story. The US-based nonprofit started in 2015. Their aspiration: To pioneer solutions to end global homelessness. New Story raises money to build houses and fosters local communities by working solely with local labor and partners. Until now, more than 2,300 homes in 4 countries and 25 communities have been funded, impacting the lives of 11,000 people. 3D printing was and is a huge new step, as over 1 billion people live without access to adequate shelter. This number is expected to rise to 3 billion by 2050. As traditional methods of building have not been able to keep up with the problem and imply huge environmental challenges, a solution like 3D printing provides a better, cheaper, and faster way of building. ⊲ newstorycharity.org

These three examples show that ideas are not only an extraordinary expression of human creativity and curiosity; tackling social and environmental challenges with a business mindset can pave the way for valuable new solutions.

TRUTH OR FICTION?

Organic plastic from algae, construction material from trash, leather alternatives from pineapple leaf fibers, clean energy from the mechanical energy of highways and railroads, robots for rescue relief, artificial leaves for converting sunlight and water accessibly and affordably into energy, deserts that can be made into fertile soil, surfaces that kill bacteria, houses built of PET bottles, lower emissions from silvopasture, robot backpacks for lifting heavy loads, mushrooms from coffee grounds, electroch-

TIP

The Katerva Awards, which spotlights the most promising new sustainability concepts in the world every year.
↗ katerva.net/home

The Red Bulletin, an Innovator magazine publication that presents startup and corporate ideas for a better future.
↗ redbull.com/us-en/innovator

Ashoka, which describes itself as "a living encyclopedia of social innovation across the world".
↗ www.ashoka.org/en-us

romic glass to reduce electricity in buildings, and caps that provide brain training: Which of these sound to you like reality and which like science fiction?

Each of these technologies already exists, and there are many more. On a dismal day, it can be very uplifting to scroll through a few popular websites that present business ideas that make a difference and whose intention is to change the world for the better. In the Tip is a list of my favorites.

Let's look at a few more innovations that make a significant contribution:

- LifeStraw, a brand that manufactures water filtration and purification devices, has designed The LifeStraw Family 2.0, a high-volume, point-of-use water filter that can purify up to 30,000 liters of water, or enough to provide a typical family of five with sanitary drinking water for three to five years. The "humanitarian entrepreneurship" approach, from parent company Vestergaard, has developed further business solutions for societal challenges, especially in poorer countries.
↗ tinyurl.com/y2b9znoy

- The Solarclave is a low-cost, solar-powered device used to safely and reliably sterilize surgical instruments in developing country clinics that lack the necessary infrastructure and tools to perform much-needed surgical procedures. It was a product designed by a team from the MIT D-Lab, whose mission is "designing for a more equitable world."
↗ tinyurl.com/y5x97rmz

- Amparo, a Berlin-based company, developed from a university project into an award-winning company. A group of students were challenged to develop a better solution for below-knee amputees. They came up with a methodology that redefines prosthetic care and has the potential to change the entire industry. Today, one of Amparo's greatest goals is to provide the 90% of amputees currently without access to prosthetic care with cost-efficient products that work even in inaccessible areas.
⊿ www.amparo.world/home-en

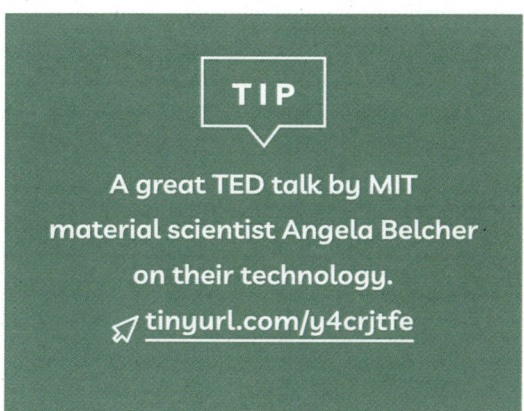

TIP

A great TED talk by MIT material scientist Angela Belcher on their technology.
⊿ tinyurl.com/y4crjtfe

- Inspired by an abalone shell, MIT scientists have found a way to program viruses to construct powerful new batteries, clean hydrogen fuels and record-breaking solar cells.

Creativity focused beyond mere profit can create great solutions and tremendous societal impact. The examples described show incredibly inspiring solutions with a value-add for people and our world. They demonstrate that innovation often has also a social focus. It is really up to entrepreneurs as well as to established corporations and research institutes to engage in this field.

6.02 —————— ARE IDEAS GROWING ON TREES?

The ideas presented up to now are of such a wide variety that they can appear random and make it easy to get confused. How do ideas come up? What are the focus areas? Which ideas are truly good ones? Let's try to get a handle on the subject of how ideas for sustainable change are found.

WILD CREATIVITY OR INTENTIONAL EXPLORATION?

Sustainable ideas can be both topics that are the focus of attention or those that are still unknown to most of the world. Many people are aware of the problem of plastic pollution in the world's oceans, in fact, most of us even contribute to this problem, but a solution is yet to be found. Affordable housing in poor countries has not yet been addressed systematically by business strategies but instead traditionally by humanitarian relief organizations. To find new solutions from a business perspective, as in the case of Vestergaard and Solarclave, is an interesting field of exploration for innovation. Many sustainable innovations aim to solve big issues that are still unresolved. Some find inspiration in unusual places, such as in nature or material science.

Other ideas focus on particular problems that often only insiders are aware of. The lack of inclusion of disabled people addressed by VerbaVoice is highly relevant when it comes to addressing a specific target group. Similarly, Amparo tackles an issue that is very present to the people affected— amputees— but seems far away to people not affected.

Our world is desperate for ideas. We need solutions for so many different problems connected to sustainability. Think about generating job opportunities, educational equity for children of difficult backgrounds, transparency about medical treatment options for patients, fighting climate change and food waste, reducing plastic in our packaging, reducing intra-city pollution, protecting biodiversity and greening value chains. Generating business models for social and environmental topics is a wonderful challenge. So, sometimes it takes a deliberate search for the right idea. Other times, it just pops up out of nowhere. Either way, the best ideas are often surprisingly simple. When we hear them, we think, "Wow, that makes sense. Why hasn't anybody else already thought of that?"

TAKING OFF FROM THE PROBLEM

Giving sustainable business ideas a framework can help to find solutions. Doing a search based on specific areas and by criteria for solutions is very common and a very good approach.

The categories of the Katerva Award, for example, are Food Security, Behavioral Change, Economy, Protected Areas, Gender Equality, Materials & Resources, Human Development, Energy & Power, Transportation, and Urban Design. ⊲ katerva.net/home
These key areas need substantial innovation in the decades to come.

TIP

See all award winning projects
from the LafargeHolcim Awards.
tinyurl.com/y36s86v6
tinyurl.com/y4crjtfe

The LafargeHolcim Awards for Sustainable Construction, the world's major competition of its kind, seeks leading projects that combine sustainable construction solutions with architectural excellence. Projects and concepts are solicited from architecture, engineering, urban planning, materials science, construction technology, and related fields. They are reviewed according to five criteria, among them are innovation and transferability, resource and environmental performance, as well as ethical standards and social inclusion. The winning projects and concepts are pioneering for the areas involved, such as architecture, construction technology and urban planning.

Finding solutions for defined problems is the typical approach to innovation in business. It can be applied to sustainable innovations, too, from startup teams to established corporations. Yet, there is a fascinating secret about innovation: Solutions are often unknown because the problem is unknown.

An eye-opener in this respect was presented to me by Latin American students in a class I was teaching on sustainable entrepreneurship. Although I had been familiar for years with sustainability and Latin America, I was not yet aware of this particular sustainability issue until they brought it to my attention as part of their everyday life in Mexico City. The students named the diaper waste as a major issue in the megacity, which back then in 2011 was the 10th largest consumer market for throw-away diapers and second in market penetration with 58%. Over 20 million inhabitants lived in the greater Mexico City area, with 2 million of them children between the ages of 0 and 4. Based on their research, the students concluded that Mexico City threw away about 1 ton of diapers every day. With waste disposal infrastructure not well established, this caused environmental and health issues. The students conducted research on how to collect and transform waste into resources and came up with three different business segments where the cellulose from diapers could be used: the paper, agriculture and plastics industries. Once a problem is identified, it often doesn't take long for people to come up with ideas. So, defining a specific problem for which a solution will later be developed is another approved approach to adressing challenges in sustainability.

 **TOOLBOX**

EXERCISE ON SUSTAINABILITY CHALLENGES

When working with students on sustainable business ideas, I would ask them to start looking for solutions to problems presented to them. This method makes people aware of the nature of social and environmental challenges from a practical perspective and trains them to think about problem analysis and market scoping. In small groups, they would deep-dive into a topic and, usually within the hour, would have developed an understanding of the problem and came up with 2-4 ideas for addressing it. By mutually presenting their results to others, they were able to learn a lot in a short amount of time.

Try it out for yourself or in a small group using the following challenges as inspiration or define some challenges relevant to you. Analyze the problem in detail and develop business ideas for solving it. Be free with your imagination, and refrain at this point from evaluating the practicability of or obstacles to implementation such as regulation, the willingness to pay, or the potential for economic losses.

– Compared to bottled water, tap water has a significantly better footprint. The water quality and infrastructure in many countries are not ready for tap water use. Plastic waste could be reduced if water were not sold bottled. How can business ideas address this challenge? Think of it from the perspective of a startup or of a company that sells bottled water.

– In retail markets, many product-related resources, such as flower-pots, bottle corks, and sales receipts, are throw-aways and designed for single-use. How can this type of resource squandering be stopped? What could close the loops? Think of it from the perspective of a retailer, an agency fighting unemployment or a packaging company.

– In cities like Hamburg, New York and Tokyo, housing is scarce and expensive. People with low income have problems paying rent or difficulty accessing any type of housing space. How could a business model address this inequality if you were an entrepreneur, a real estate company or a hotel owner?

– Pharmaceutical companies go through long and expensive research and clinical trials in order to release a new medication, only then to receive patent protection for about twenty years. This makes the incentive too low to encourage companies to conduct research on diseases that only effect a small number of people (called orphan diseases) or diseases prevalent in poor countries. How could this be addressed by an entrepreneur? And by a pharmaceutical corporation?

STRAIGHT FROM THE HEART

People love to create and to innovate. Businesses do this all the time. In sustainability, it's less about finding the next technological novelty, a new fragrance, a special anything, so people have just more fun or buy more. The primary motivation is not to raise market shares or generate profits as an end in itself but to provide solutions, as many examples reveal.

Innovators in sustainability are often motivated from the heart to create solutions that have an impact beyond themselves and the money generated. Or they may feel an urgent, pressing need to see a change in an issue.

Ocean Cleanup founder Boyan Slat believes that it is simply not acceptable to have plastic masses floating around out there in the oceans. He shares that motivation with Parley for the Oceans founder Cyrill Gutsch who attacks the issue from another front by working with companies to change the apparel industry (see page 167).

Zafer Elçik from Turkey was called a social innovation hero by the Red Bulletin. He had an issue that was close to his heart from a personal experience that he ended up translating into a business: *"When Zafer noticed the only thing that captured his autistic brother's attention for any time was his smartphone, he built a gaming app, hoping to enhance his brother's cognitive skills. His project is now Otsimo, a free, Turkey-based educational gaming app, helping more than 30,000 children with learning disorders and special needs to improve cognitive learning. ... When I first started out, I didn't do it for the sake of creating a business, I felt a strong desire to help my brother. Only then did I realize my idea had great potential to help others— Otsimo grew organically into a business. Our mission now is to democratize special education with the help of technology."*[98]

Countless innovators begin in just that way—they are deeply touched by an issue, personally impacted or feel an urge to change the status quo. It's a truth about sustainable innovation: A powerful source is released when people develop a solution for an issue they perceive as really near to their heart.

This kind of motivation can be modeled in a lab atmosphere. Contrary to the conventional approaches taught in big corporation innovation workshops, at this stage, it would not be about solving a specific problem but about accessing where a person's heart is involved. What hurts or moves a person in daily life or his or her social surroundings? Which of the several global issues grab's the person's heart? Why do they want to see change? What is close to people's hearts is very individual.

This was usually the second step I took with the students after the one described in the toolbox. When students were asked to search for a heartfelt issue, they came up with very personal ideas. It was always a moving and surprising moment when they presented their issues and ideas. They often reflected their own social surroundings, which were naturally very different for people in Germany than in places such as Mexico City. From there, students developed business plans. Over the past several years, there has been a surge in business plans tackling a number of issues:

- Bottled water is usually sourced far away, involving extensive transport. How could an international water producer build up local water sources and retail networks? How could a local city water brand be developed?

- Babies outgrow clothes very quickly. What could a business model for an established retailer look like that would allow bartering for baby clothes, in the form of perhaps a subscription model that grows with the baby?

- In Colombia, voluntary initiatives for carbon storage through forest projects are not widely adopted, and land use rights are often violated. How could incentives for avoiding deforestation be put in place and at the same time, income sources for small farmer communities be created?

- Bananas, cucumbers and others have to have a certain size and shape under EU regulation. How could a retailer promote diversity among "imperfect" fruits and vegetables to avoid food waste?

- People in cities like to live more consciously. How can they be inspired to create their own upcycling crafts?

Most established corporations don't create the freedom and space necessary for people to open up about issues close to their heart and develop solutions for them. I dare to state that this is one reason why many sustainability ideas originate from startup entrepreneurs.

DREAMING THE IMPOSSIBLE

In 1899, Charles Duell, chief of the US American patent office, purportedly said: Everything that could possibly be invented has been invented already. This was at a time when the car and radio had just come onto the market. Some decades later, in 1943, it was predicted that the worldwide demand for computers would be: five.

It's a truth about ideas and innovation: A solution is often inconceivable only because the imagination is too limited.

Some people—usually entrepreneurs but also scientists, artists or activists—see things before others do. Or they are bold enough to dream that things could be different. They question the status quo, think of something new and better. They see opportunities where others see either no problem at all or are convinced that things can't be changed, something to the effect of "things have always been this way". Often, the forerunners fight against obstacles and failure over long periods of time to effect change and progress.

This is true for big business inventions. Read, for example, about the history of today's civil aviation. Many people over centuries worked on the dream to fly. But it also holds true for social and environmental innovation. For example, William Wilberforce dreamt of a world free of slavery and spent many years leading the fight to abolish slavery in the British Empire. Felix Finkbeiner, a nine-year-old fourth grader from Germany, started the initiative Plant-for-the-Planet in 2007. Three years later, already one million trees had been planted. These people dreamed of something others didn't see.

TIP

A recent Deloitte report looks at "Is technology an enabler for sustainability? How blockchain can be used across value chains to achieve sustainability".
tinyurl.com/yxtsen9h

Can you imagine a world without computers today? In 1943, and for many years thereafter, it was inconceivable that computers would become relevant in daily life. Someone had to imagine it, research it, develop it, and get it prototyped. When transistors and microprocessors were invented, the technology could enter into the mass market. Only at a certain point, a solution becomes advanced, self-evident and scalable, if it reaches that point at all. And then, inventions often lead to a chain of new inventions. This is true for business innovation as well as for ideas in sustainability. This is why it is important to encourage and support sustainable innovations—especially when it comes to dreaming the impossible.

In truly genuine ideas, there is always also a spur of hope. Results cannot be planned. But boldness is probably an attitude. Mark Twain is attributed with having said: "They did not know it was impossible, so they did it."

FROM EMERGING DAVIDS AND GREENING GOLIATHS

Most of the ideas presented so far come from startups; some of them already at a more advanced stage. Coincidence? Likely not. Many ideas in sustainability until now have emerged from startups. Big corporations often focus far more on scaling up ideas rather than inventing them in the first place.

DAVID & GOLIATH IN ECONOMIC THEORY

A powerful illustration of this market dynamic is described in a concept that has its roots in a scientific paper from 2010 by the economists Kai Hockerts and Rolf Wüstenhagen. This paper also provides a good overview of the history of research in the field of innovation and sustainability.

TIP

The story of how underdog David conquered the overpowering Goliath stems from the bible, 1 Samuel 17.

⌁ tinyurl.com/yyp2nrse

The authors describe small firms, or startups ("Emerging Davids"), and established corporations ("Greening Goliaths"), which can be distinguished by their sustainability ambitions and their market influence. Both contribute to the transformation of industries towards sustainable development: "... *in the early stages of an industry's sustainability transformation, new entrants ('Emerging Davids') are more likely than incumbents to pursue sustainability-related opportunities. Incumbents react to the activities of new entrants by engaging in corporate sustainable entrepreneurship activities. While these 'Greening Goliaths' are often less ambitious in their environmental and social goals, they may have a broader reach due to their established market presence. This paper analyses the interplay between 'Greening Goliaths' and 'Emerging Davids' and theorizes about how it is their compounded impact that promotes the sustainable transformation of industries ... We argue that in the early stages of an industry's transformation towards sustainability, it is typically small firms and new entrants that stimulate disruptive sustainability innovation. Attracted by the early market success of Davids, Pioneer Goliaths follow up with corporate sustainability entrepreneurship initiatives of their own. Thanks to their larger scope, these initiatives take the sustainable transformation of an industry to the next level.*" [99] The following graph illustrates the dynamic:

Sustanability market transformation

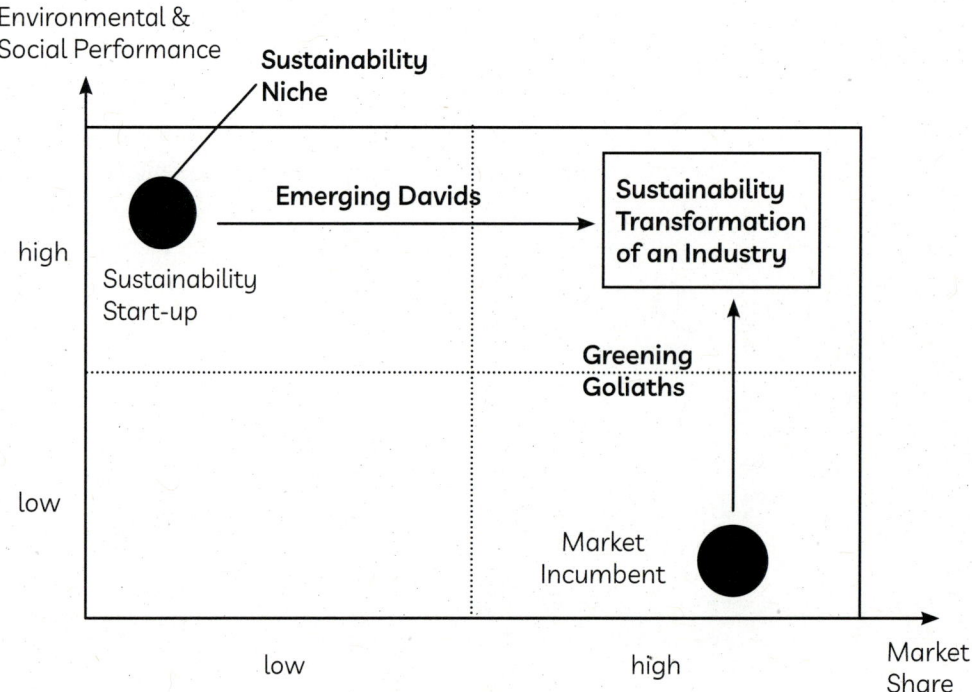

Based on source: Hockerts, K., Wüstenhagen, R. Greening Goliaths Versus Emerging Davids—Theorizing about the Role of Incumbents and New Entrants in Sustainable Entrepreneurship. Journal of Business Venturing 25 (2010) 481-492. With permission of Elsevier.

CREATING A SUSTAINABLE REALITY

The described dynamics have taken place in several industries over the past 10 to 20 years. One very obvious example is the food industry, including retailers as important gatekeepers in the value chain (see pages 82f.).

Numerous startups in the food industry have brought trends like organic, Fairtrade, vegan, regional, no additives, and less packaging—just to name a few. Established corporations, many of them multinational brand companies, have slowly but increasingly adapted to these trends. As the market approached a certain level of development, big corporations sped up, adapted existing solutions and scaled up. Retailers opened the gates to more sustainable products and embarked themselves into more sustainability. This, in turn, invited new startups and pushed established companies further. Today, the level of sustainability in the food industry in terms of variety and availability is much higher than in the early 2000's. Similar developments can be observed, for example, in the cosmetics, home cleaning and textile industries.

Also, in technology-driven areas, it can be observed—the emergence of renewable energies and electric cars mainly was driven by newcomers. Established corporations picked up much later. Also, most innovations in the social sector are meanwhile pursued by startups. Established corporations mostly limit themselves to community contributions (see chapter 8) rather than addressing social issues with a business approach as Vestergaard does. Although there is a fine line in combining societal needs with business approaches, it will be interesting to see if the dynamic will change over the next decades or if the role of established corporations will continue to be mainly in scaling up, if at all.

TIP

In 2009, I co-authored a book on sustainable innovations in the food industry comparing sustainability-oriented startups with established corporations innovating in a new arena.
This book is available in German through publishing company Behr's: "Nachhaltigkeits-Innovationen in der Ernährungs-wirtschaft",
Authors: F.-Th. Gottwald/ A. Steinbach

— **KICKSTARTING SUSTAINABLE INNOVATIONS**

From my background as an economist, entrepreneur and strategy-focused sustainability consultant, innovation and market development always held a special fascination for me. I should first emphasize that in sustainable entrepreneurship and sustainable innovation, the fundamental logic of entrepreneurship and innovation holds true. The right moment is key, just as much for a startup as for growing firms. Some ideas succeed, some fail.

There are, however, several specifics in sustainable innovation due to the complex nature of sustainability and a usually underdeveloped market setting. Some key aspects will be marked out for startups (entrepreneurs). Let me also share some observations of the setting for established corporations (intrapreneurs), which in many areas is still more greenfield.

SCALING UP AS SUSTAINABLE ENTREPRENEURS

In 2020, mymuesli celebrated its thirteenth anniversary and with that surpassed the critical 3-to-5-year span decisive for the survival of a startup. It's one of those stories we love about entrepreneurs. In the summer of 2005, three young friends from southern Germany were at a local beach where they came up with the idea to have people create their own individual organic muesli. They set up an online startup, mixed the muesli in their home kitchens and spent a long time packing the parcels themselves. Today their goal is to make the best muesli in the world but make it sustainable in all ways possible. The company has set standards as to what is allowed for its product range: 100% organic ingredients, no palm oil, German-manufactured, carbon-neutral delivery, improved packaging, high transparency and customer involvement when it comes to the status quo and challenges where the company wants to get better. Website in English: ✈ uk.mymuesli.com

We saw that startups are prominent in sustainability. They are not lacking in inspiration, the power to dream big, or the desire to find challenges and solutions. Often startups are like scouts for sustainable business ideas.

A good business plan is mandatory for each startup, no matter what business focus it has. The personality, drive, and perseverance of the innovator are decisive. A nourishing structure, an entrepreneurial atmosphere, as well as good partnerships, foster success. Entrepreneurs with a societal focus share the same problems with all other entrepreneurs: They often have too little money to start up or scale up. They need to be allrounders yet lack certain skills and have too many tasks to solve at the same time. Growth barriers have to be overcome.

Other specific challenges facing sustainable entrepreneurs making it even harder can be found, in my view, in the following four areas:

– Carving out the social issue: A sustainable startup needs a business plan that reflects, in detail, the social or environmental challenge it wants to address. It must deeply understand the challenge's roots, dynamics and stakeholders involved, as well as the anticipated developments and broader competitive environment, and describe strategies for addressing these. This perspective has to be continuously adapted to new developments as changes in the framework can occur at a rapid pace, and competitors may catch up quickly.

– Underdeveloped markets: The sustainable startup often enters a complex arena where many of the basic conditions hinder development. Value chains do not yet exist, need to be built up or fundamentally changed. Groundbreaking research has to be done. The willingness to pay must be created. Policy regulations have to be formed, and so on. The challenges are very specific to the innovations. All this has to be taken into account in terms of time horizons, resources (skills, money) and focus. Specific help in addressing these challenges has to be organized, for example, in the form of practical knowledge through alliances, business angels or other sources.

– Overcoming excessively high standards: Sustainable entrepreneurs often aspire to create business models that are 100% perfectly sustainable and at the same time achieve the highest standards in all aspects. This noble aspiration can be backbreaking for the entrepreneur, especially if it is nurtured by demands from customers and the public. Sustainable businesses will very likely encounter barriers at some point and have to accept trade-offs, as well as decide if it is economically responsible to achieve a certain improvement. Companies are recommended to be very transparent about their aspirations and limitations and let the public know what can and cannot be done. mymuesli is a good example of this in its ability to engage with customers and the public about the current limits to sustainable packaging and involve them in finding better solutions.

– Find the right growth momentum: At some point in the growth process, new key decisions on sustainability come up. Usually, these decisions stem from the need to balance the social or environmental issue and economic considerations in a new way. Sometimes, a successful brand is faced with a takeover, or it must find a suitable investor. Or, as in the example of a sustainable food brand I worked with, must make a decision as to whether it wants to grow while accepting a less favorable footprint, for example, through energy-intensive production processes or more packaging. Reflecting on decisions with a good sounding board and mapping out options can be useful approaches in such situations.

OVERCOMING INERTIA IN BIG COMPANIES

Katerva (see page 180) structures sustainable innovation in established corporations along four levels, covering the spectrum from incremental to radical innovation.

– Level 1 incremental: Incremental or small, progressive improvements to existing products

– Level 2 re-design or "green limits": Major re-design of existing products, but limited the level of improvement that is technically feasible

– Level 3 functional or 'product alternatives': New product or service concepts to satisfy the same functional need e.g., teleconferencing as an alternative to travel

– Level 4 systems: Design for a sustainable society[100]

The focus of today's sustainable innovations at established companies is—if it at all—on levels 1 and 2, with a few examples on level 3. The most relevant innovations in the future from the corporate side will likely be required from levels 3 and 4, which are also the focus of this chapter on innovation. Accomplishing this will require a new degree of boldness and likely new approaches at corporations.

Yunus Social Business (see page 224) did an extensive study on sustainable intrapreneurship, called "Business as Unusual". They found out that the majority of initiatives operated as a project of a specific department. The portfolio analyzed a great variety of issues, from youth empowerment to mobility, agriculture, housing to circular economy and malnutrition. The top three initiatives of sustainable intrapreneurship were in the fields of healthcare, accelerators/boot camps and water & sanitation.[101] It will be another challenge for huge corporations to engage in innovation at a larger scale than for projects or single departments.

Established corporations possess resources such as knowledge, people, money and structures for innovation, and they often have the market power to scale up innovation. In the model above, "Greening Goliaths" are said not to have been the driving force behind sustainable innovations in the early stage. This reflects my experience after working for over 18 years with big corporations on sustainability. Most companies focus on managing sustainability, limiting risks and greening their existing businesses. In addition to planning and executing innovation in sustainability, developing suitable ideas is among the most challenging tasks for established corporations. Realizing sustainability business opportunities by thinking systematically, starting from the current business model is scarce.

Several big companies have set up accelerators or startup programs to spark entrepreneurial spirit and develop and in-source new ideas. Only a few of these companies, however, systemically include sustainability or place a focus on it. At the same time, the typical problems of innovation in big companies and challenges arising from sustainability management ultimately apply to these efforts. Responsibilities for sustainability projects are insufficient as they often occur outside of typical organizational boundaries. Opportunities and trends are not sufficiently understood or at the radar due to lack of know-how. Often, they are not addressed early enough. The necessary courage and ownership for long-term projects with high uncertainty and complexity are lacking. The day-to-day routine as well as incentive focus on classic margins instead of future revenues. Solving sustainability challenges is not encouraged; rather the next product variation of a well-selling product will be pushed. The purpose and direction of innovation are deeply installed and not questioned—and sustainability is not implanted into it. And sometimes, a sense for the fine line between courage and being "over the top" with ideas in terms of their approach and marketing is missing due to a lack of experience or an underestimation of the special nature of sustainability. All of this prohibits effective, sustainable intrapreneurship in companies.

Let me share some anonymous client examples from my business experience.

– The research and development leader of a high-tech manufacturing company focused on engineering excellence and doubted that environmental aspects would be of any relevance to the product portfolio. Despite examples from competitors and evident dynamics from other industries, he neglected to set up a project with developers for early analysis and prototyping. Some years later, the industry dynamics had accelerated, and increased energy efficiency had become a major requirement of customers. The company had not been able to leverage its technical excellence to the next level by integrating sustainability.

– An innovative company in packaging had advanced for years in sustainability and introduced packaging to the market with improved footprints. The next bold step would have been to actively engage in co-creation with key customers from key industries. This would have allowed the company to actively shape the future demands for sustainable packaging that could already be seen on the horizon. Focus on protecting existing margin projects and lacking the courage to invest in new approaches within the value chain had closed the window of opportunity.

– Together with an international multi-brand consumer goods company, several ideas for new sustainable concepts in products were developed in creative processes. A fear of endangering the established brands led to the choice not to implement ideas but instead stick with the traditional products that had a low sustainability profile.

– A food service company was eager to implement sustainability and developed approaches to differentiate itself in the eyes of the customer using new sustainability concepts. Higher costs on the supply side and disappointing initial marketing experiences caused the projects to shut down prematurely.

– Together with an advanced industrial technology company, trends and opportunities were developed for future business activities in sustainability. Getting bogged down in day-to-day operations and a lack of commitment from the CEO led to not pursuing ideas further.

– The analysis of a leading chemical company showed that innovation processes and structures had not been set up for sustainable innovations. For example, sustainability innovations were not able to pass through the typical gate processes as complexity and business returns in the initial stages are much different than what a large corporation is accustomed to with the typical innovation projects. At the same time, the search fields and business focus were not to be on sustainability challenges and opportunities. Another setup for processes and structures was required and successfully implemented.

This list is not exhaustive but reflects how challenging it can be for established corporations to address sustainable innovations. But, as sustainability is a trend that is here to stay, it is vital to invest in sustainable intrapreneurship.

What could be a solution? Other than addressing the organizational barriers described, a great starting point would be to search systematically for ideas for sustainable innovation dynamics. I suggest three measures that can work effectively.

 **TOOLBOX**

THREE WAYS FOR KICKSTARTING SUSTAINABLE INTRAPRENEURSHIP

1. Conduct a yearly review on the status of sustainable innovation readiness, including an analysis on strategy, competitive environment, major trends and best practices, as well as internal processes and structures.

2. Establish an annual or bi-annual workshop on ideas for sustainable businesses with a cross-functional and cross-divisional team, and specifically including young people for imaginations without boundaries, working on specific sustainability challenges for the business but also allowing time and space for the heartfelt issues of the employees involved.

3. Invite external people for challenger sessions on upcoming or existing ideas towards sustainability.

These measures can kickstart or, conducted repeatedly, nurture sustainable intrapreneurship. They would help to keep the topic on the radar and implement sustainable intrapreneurship into business routines. This is especially relevant as it is diffcult to implement it ad-hoc when necessary structures and mindsets are not oriented enough towards sustainable innovation, as described in the passage above.

MAKING IT WORK

At the end of this chapter, I would like to share first five successful entrepreneurship examples from people in Hamburg. These examples demonstrate a wide variety of ways to add value and address challenges using a business approach.

– Viva con Agua: This network of organizations and people engages in different activities to provide worldwide access to water. Products are one way to generate income for projects and create awareness in everyday life for issues. The Viva con Agua mineral water is meanwhile a well-known favorite brand in retail shops and restaurants. Despite selling bottled water, Viva con Agua is a strong advocate of drinking tap water as the most sustainable way to consume water in Germany. Access to water projects are financed with the revenues. One of its latest products is toilet paper called "Goldeimer" (golden bucket). It is made out of 100% recycling paper and brazenly marketed as "No trees for asses". Its cool design raises awareness of the fact that 4.2 billion people don't have access to sanitary facilities. Goldeimer spends 100% of its profits on building toilets and raising awareness of this issue by touring with compost toilets to festivals. Instagram: @vivaconagua-worldwide

– SofaConcerts: This company, founded by two young women, built a platform for artists to play private, live home concerts. SofaConcerts supports social get togethers through music and creates opportunities for artists to gain an audience and income. They now have over 4,000 artists on their platform. During the Covid-19 crisis, SofaConcerts came early on up with the idea to support the musicians who were severely hit by the pandemic. SofaConcerts launched a campaign whereby musicians created private videos, where they would sing songs and send private messages to loved ones, thereby bridging social distancing while generating revenues at the same time. www.sofaconcerts.org/en

– Avocadostore: Created in 2010, Avocadostore was the first platform for green products and is today Germany's biggest marketplace for eco-fashion and green lifestyle with 200,000 products from 4,000 brands. Avocadostore created a list of sustainability criteria that products have to fulfill. They provide transparency in the field of sustainable products and broader access to sustainable startup brands for customers, e.g., in eco-fashion.

This is also actually a viable model for online retailing as they ship parcels carbon-neutral, and most companies prefer limited and natural packaging, such as cardboards and avoiding plastic. ⟋ www.avocadostore.de

TIP

The book "Drawdown" (edited by Paul Hawken) became a New York Times bestseller in its first week of publication. Bringing together geologists, engineers, agronomists, climatologists, biologists, botanists, economists, financial analysts, architects, NGOs, activists, and other experts, "Drawdown" offers 100 solutions to reverse global warming, many of them from corporations and involving an abundance of technology.

– Dialogue in the Dark: In 1988, Dr. Andreas Heinecke started a social business, working with blind people. Blind people do not have equal access to education and the labor market. In addition, there is widespread prejudice about blindness in society. The concept Dialogue in the Dark enables people to encounter blindness through exhibitions and events such as dinners and business workshops held in the dark. Today, it is an international network present in more than 40 countries with different formats adapted to the local cultural context. More than 9 million visitors have participated, and thousands of blind guides and facilitators find employment through exhibitions and workshops. The concept was extended to other disabilities such as deafness with "Dialogue in silence". ⟋ www.dialogue-se.com

– BIO-LUTIONS: This company offers sustainable packaging and disposable tableware solutions made of agricultural residues. Natural raw materials are converted into self-binding natural fibers through a mechanical process. The characteristics of the natural raw materials offer an excellent footprint and an alternative to plastic packaging. Bio-Lutions creates a decentralized production network with local factories and regionalized distribution, as well as localized raw material collection and processing, which is also an important factor for competitive pricing. The goal is to operate 40 factories in 40 countries by 2024. ⟋ www.bio-lutions.com

Let's at the end also look at three established companies where business opportunities stemming from sustainability were realized despite encountering challenges along the way. These are typical level 3 and level 4 innovations and point out typical obstacles.

– Rügenwalder Mühle is a German manufacturer of sausages and cold cuts, founded as a small butcher shop in 1834 and employing 600 people today. In 2012, the company started a major transformation towards vegetarian products. The rationale: if the world population grows as predicted, meat production will double worldwide—something for which our planet has not enough water, feed or space. The company dared to take the bold step of offering nuggets and burgers made of soy and wheat, as well as a wide range of vegetarian cold cuts and sausages, right next to its range of meat. It was a long road to convincing people internally and externally, including retail partners and customers, in order to gently evolve the brand Rügenwalder Mühle. Today, 25% of the company's revenues are generated by the vegetarian and vegan product range, with the goal to raise this in the near future to 40%.[102] A critical article highlights that the turnover of Rügenwalder's sausage business has only slightly decreased since the launch of the vegetarian and vegan product range. Thus, the overall business expands, but no significant effect on the reduction of meat consumption is achieved. Vegan advocates blaim the Rügenwald example as greenwashing.[103]

⬦ www.ruegenwalder.de/en

– Danone formed a social business, Grameen Danone Foods, that strived to connect business opportunities with a social value-add. Through this company, Danone introduced fortified yogurt as a dose of daily healthy nutrition to low income, nutritionally deprived populations in Bangladesh. It also strengthened local dairy farmers and increased local job opportunities. Grameen Danone Foods remained challenging as a business case, especially due to

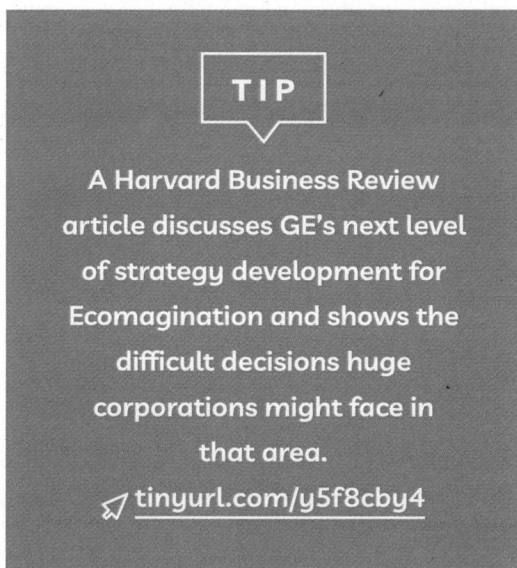

TIP

A Harvard Business Review article discusses GE's next level of strategy development for Ecomagination and shows the difficult decisions huge corporations might face in that area.

⬦ tinyurl.com/y5f8cby4

TIP

@sustainabilitychampions presents on Instagram regularly inspiring ideas and people in a broad panorama of sustaina- bitliy. Again, currently, most of them are entrepreneuers.

the difficult local conditions, such as maintaining cooling chains. Out of this venture rose a focus in Danone's 2030 strategy to create inclusive economic growth, among other things, through empowering vulnerable communities linked to its value chain, including farmers, waste pickers, street vendors, and caretakers.

More about this can be found at
⚲ tinyurl.com/yyr8p5wm

– General Electric (GE), in 2005, launched back then a groundbreaking strategy early on with Ecomagination to build more efficient machines that produce cleaner energy, reduce greenhouse gas emissions, clean water and cut its use, and make money while doing it. By 2015, Ecomagination had generated more than 200 billion US dollars in revenue since its inception. After that, a fine line was tested by GE's expansion of the program through eight partnerships to accelerate innovation in water and energy efficiency. The company committed to investing 10 billion US dollars in R&D in cleantech technology by 2020. Among its priorities were alternative water technologies for fracking natural gas, a practice seen as highly controversial by environmentalists and causing fundamental challenges. Today, GE has integrated the Ecomagination approach and its projects into various divisions such as oil and gas and healthcare.

TOOLBOX

SEVEN CRITERIA FOR SUCCESSFUL INTRAPRENEURSHIP

I suggest it can be said that successful sustainable innovation projects within established companies comprise at least several of the following criteria:

1. They are a strong fit with the core business. They grow over time in credibility and acceptance.

2. They are novel and develop solutions outside of established business boundaries, addressing a truly relevant social or environmental challenge.

3. They often partner with others, for example, research institutes and other companies.

4. They usually have a clear commitment from the executive board.

5. They successfully overcome organizational barriers to innovation.

6. They involve a team of highly engaged leaders and employees.

7. They scale up at a certain point. And they subsequently impact the core business to change towards more sustainable paths.

APPLICATION

1. ———— Think about a problem at your company,
in your direct surroundings, or in society that
you think needs to change.
– State the problem.
– Research it.
– Join some friends or colleagues to
come up with three business ideas that
could help address this problem.

2. ———— Think about the products or services of a business
that you really admire, regardless of whether this is
considered a "sustainable business" or a regular,
conventional business.
– Why do you admire it? What are they getting "right"?
– Which three to five key success factors
make it successful?
– Is there anything the company does that can be
applied to the problem you stated in No. 1?

3. ——— If you work in a corporation:
– How does your innovation agenda look?
– Does your company explicitly pursue solutions
to societal challenges and make this a high priority?
– If you think there is room for improvement,
what could you do personally to trigger change?

4. ——— Which in your view have been the most
relevant innovations of mankind? Why?

5. ——— In relation to sustainability: What do you
– think is impossible to invent?
– want to see invented in the next 20 years?
– want to see invented in the next 50 years?

THE PEOPLE FACTOR

For the past several years, I have been deep-diving into the topic of new work as part of my voluntary job at our church. I have talked to many people about their jobs, read reports and books about the future of work and created with our team projects around new work from a faith perspective. So, my long-year passion (sustainability) was joined by a very practical one (work).

These reflections, together with ongoing client work, revealed: Sustainability directly interferes and interacts with the job world in multiple ways, and new work triggers sustainability in a special way. Moreover, I believe work is the "new sustainability", a next megatrend reshaping our businesses and economies. It is remarkable to already recognize similar developments like in the early days of sustainability. And as people are much more directly affected by work, compared to sustainability, developments will have a tremendous speed. We will see a massive transformation, and it is on us to shape it well.

THE FUTURE OF WORK

In the last few years, several hotspots have emerged about the value and meaning of work. Among them are questions like how technology changes work, how should the opportunities created from remote or agile work be shaped, and how should jobs be designed in the future. Other questions include how we deal with work-related problems in society, such as un-employment, a lack of skilled workers and eroding social security systems. Demographic change, digital divide, educational and income disparities and lifelong learning, as well as the provision of decent jobs, are among the challenges ahead. Covid-19 spurred remote working.

Questions arise at different levels—an individual, company, industry, regional, country and global level, to name a few. They are addressed in public, the media, the arts, and scientific communities. Numerous influential organizations have published reports or launched initiatives on the future of work, among them the International Labour Organization (ILO), the United Nations, the Organization for Economic Development and Co-operation (OECD) and the World Economic Forum (WEF), consultancies like Deloitte and McKinsey or companies like Microsoft.

THE HELICOPTER PERSPECTIVE

The new work discourse is often connected to an individual, a company or an industry, especially in richer countries. It typically refers to the challenges and opportunities arising from technology and people's stronger desire for self-realization and purpose.

But let´s start with some trends at the global and macroeconomic levels because, in the long term, these will cause significant interference between people and sustainability.

The creation of decent work opportunities will be a top challenge for our global future, mainly as a result of demographic change. The current world population of 7.6 bn people is expected to reach 8.6 bn in 2030 and

TIP

The OECD website provides analysis on demography and job projections at country level, self-tests, stories and recommendations for the future of work.

⚐ oecd.org/future-of-work

9.8 bn in 2050. According to World Bank estimates, over 600 million new jobs will need to be created by 2030—equivalent to around 40 million jobs per year—just to keep pace with the growth of the working-age population.[104]

In some regions, such as Germany, the population is shrinking and getting older. This puts pressure on social security systems and the overall economy. In other regions, the population is rising enormously. By 2050, the African continent is supposed to count for half of the worldwide population and a high percentage of young people. This growth means increased pressure on natural resources for meeting basic needs, which directly leads to a surge in the importance of sustainability. At the same time, there is a need to provide income and work opportunities for a growing number of people. Major concerns are unemployment and work that doesn't pay enough to cover the cost of living.

According to the UN, the global unemployment rate stood at 5% (2019). At 11% unemployment, joblessness is particularly pervasive in northern Africa and western Asia. The unemployment rate is considerably higher among young workers than among adults in all regions.[105]

The ILO refers to working poverty, where about 780 million women and men are working but not earning enough to enable them and their families to rise out of two-dollar-a-day poverty.[106] The UN calculated that in 2016, 61% of workers globally were engaged in informal employment, lacking formal contracts, especially in the agricultural sector (94%). Informality poses challenges with regard to earnings, work schedules, occupational health and safety, and working conditions generally—crises like Covid-19 hit informally employed workers especially hard.[105]

Richer countries are also faced with this issue: One in seven workers in OECD countries is self-employed, working part-time or under a temporary contract. Currently, 35% of the US workforce is in supplemental, temporary, project, or contract-based work. The freelance workforce is expanding faster than the total workforce, with growth of 8.1% compared to 2.6% for all employees.[107] All of these workers are supported by a very precarious safety net with respect to unemployment help, job security, and access to health insurance and pension plans.

TIP

A young German social entrepreneur developed a system which involves customers to pay fairer wages to producers in developing countries.

⤳ www.tip-me.org

All this relates to the questions of whether or not people can earn enough from work to make a living and are they working under conditions that we consider dignified. Decent work is therefore one of the 17 Sustainable Development Goals. Promoting jobs and enterprises, guaranteeing rights at work, extending social protection and promoting social dialogue are the four pillars of the ILO Decent Work Agenda, with gender as a cross-cutting theme. The link between decent jobs, economic growth and advancing the development agenda is seen as crucial.

The ILO analyzed in 2016 how sustainability and decent work correlate in data across the world. They drew a sobering conclusion: *"Few countries combine an environmentally sustainable footprint with decent work. The key issue for the future of work is that nearly all countries which are sustainable from the point of view of the environment, have very high rates of working poverty. Previous development models suggest that moving these people out of poverty necessitates an increase in the ecological footprint. But this will result in environmental degradation that is also likely to destroy jobs and incomes, with the impact most felt among particular groups (e.g. migrants forced from their home by climate disasters, indigenous and tribal peoples, people with disabilities, the poor). In some cases, environmental degradation may also aggravate gender inequality."*[108]

The future of work must entail an environmentally sustainable development path. Economic growth should target the creation of decent work

TIP

An ILO report on decent work and the 2030 Agenda give a good overview of work in relation to the SDGs.
tinyurl.com/y3oheefy

while considering the environment and its natural resource base.

Inherently, it should also consider the value of work in itself in all of its facets, as was stated in a 2015 report of the United Nations on human development: *"Twenty-five years ago the first Human Development Report in 1990 began with a simple notion: that development is about enlarging people's choices—focusing broadly on the richness of human lives rather than narrowly on the richness of economies. Work is a major foundation for both the richness of economies and the richness of human lives but has tended to be conceptualized in economic terms rather than in human development terms. The 2015 Human Development Report goes beyond that convention in directly linking work to the richness of human lives. This Report starts with a fundamental question—how can work enhance human development? The Report takes a broad view of work, going beyond jobs and taking into account such activities as unpaid care work, voluntary work and creative work—all of which contribute to the richness of human lives."*[109]

HOW TECHNOLOGY CHANGES WORK

Much of the discussion around the future of work circles around technology and its impact on jobs. Major transformations occur as a result of technological change: industrial automation and robotics, artificial intelligence, augmented reality, 3D printing and other disruptive technologies are already introducing radical changes to workplaces and whole industries. Technological change, as well as demographic and socioeconomic change, are seen as paramount drivers of change for the future of jobs, according to a 2016 study of the World Economic Forum. In the demographic and socioeconomic sphere, most respondents rate the changing nature of work and flexible work as the main drivers (44%), far higher, for example, than geopolitical volatility (21%), rapid urbanization (8%) and even young demographics in emerging markets (13%). In the technological sphere, mobile internet and cloud technology (34%) lead the way; other technological aspects such as the Internet of Things (IoT), the sharing economy and crowdsourcing, artificial intelligence, robotics and autonomous transport, advanced manufacturing and 3D printing and advanced materials/biotechnology are also mentioned but by only 6-14% of respondents.[110]

TIP

"The Future of Job Report" (2016) analyzes dynamics overall and for industries and countries.
tinyurl.com/jzlr5le

"The future of work: is your job safe?" (2019) is a good documentary by The Economist.
tinyurl.com/yxwha3qw

According to a 2018 World Economic Forum report, it is projected that while nearly 1 million jobs may be lost, another 1.75 million will be gained: The jobs of the future are expected to be more machine-powered and data-driven than in the past, but they will also likely require human skills in areas such as problem-solving, communication, listening, interpretation, and design.[111]

Beyond the predictions concerning massive job losses or opportunities for job gains, I would like to emphasize three points.

Number one: It is up to each of us as citizens and as a society as a whole to discuss and decide how we want to live and work in the future. Technology offers great opportunities but it should not rule us. We can shape and manage technology such as industrial automatization. We should not leave that space solely to economic considerations within companies and industries. Instead of "what's possible?", a guiding thought should be: "what serves people?" A broader public discourse about likely developments and different paths would be desirable. Ultimately, this is also an area for policymaking, as can be seen from the efforts of multilateral organizations such as the aforementioned ILO and UN, and of the EU.

Number two: There is a need for digital responsibility, for example, in terms of data security and managing dynamics of corporate power on the internet. These vast fields can only be discussed with many actors contri-

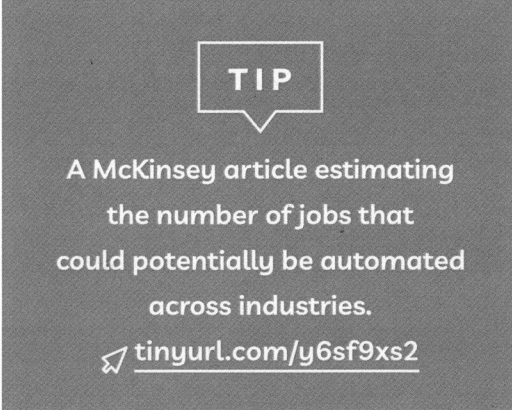

TIP

A McKinsey article estimating the number of jobs that could potentially be automated across industries.

⤳ tinyurl.com/y6sf9xs2

buting to the debate. To deep-dive further, I would like to mention an extensive study of Bertelsmann Stiftung on corporate digital responsibility (2020, available only in German) with a multi-faceted debate. As an example of a concrete action, I would like to mention the initiative of Deutsche Telekom #TAKEPART—No hate speech.

It specifically addresses hate on the internet as a societal problem. Thus, it is crucial to take a stance and strengthen digital civil courage. ⤳ tinyurl.com/y6bqmvuq

Number three: If we strive for technological transformations such as massive automatization and artificial intelligence, we must take responsibility as societies and citizens for its consequences. For example, according to the OECD, six out of ten workers do not possess basic computer skills.

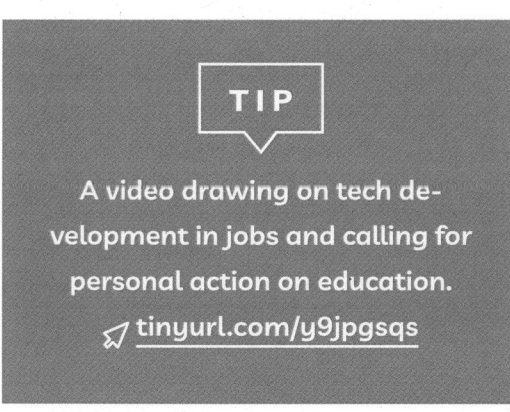

TIP

A video drawing on tech development in jobs and calling for personal action on education.

⤳ tinyurl.com/y9jpgsqs

With technological developments on the way and the expected disruptions of workplaces, this is no longer just a personal issue but a societal one. How can we as societies create job opportunities for lower skilled people? How do we truly help people to keep up with technological developments and gain the skills needed? How do we deal with people who have been left behind? Does it matter to us whether there are enough decent jobs around?

This is a broad area of involvement for companies just as much as it is for citizens, and it will need to increase in the future. I am convinced that decent work opportunities for everyone will not appear on their own but will have to be made a priority not only by politicians but also societies, companies, consumers and everyday citizens. We will likely even have to re-open the discussion one step earlier: Do we truly agree on the goal to provide decent work for all? Should we make it a priority? Are there alternatives? We must increasingly see economic developments and people's concern in their broader context and in relation to the vision of how we want to live. That's one thing that work has in common with sustainability.

MATCHING PEOPLE & SUSTAINABILITY

The original definition of sustainability described people as being integral to caring about equal opportunities for future generations (see page 46). Within the concept of sustainability, one of the three pillars is "social", which leads to a social dimension included in management and reporting.

But how does it translate and connect to people? I conducted an analysis for a client about how the social dimension breaks down into "people" for the specific context. Though it varies according to the industry a company operates in, four major lines of interference between people and sustainability can be drawn.

– The employer perspective: This comprises how a company treats its own employees, cares for their safety, health and wellbeing, their personal and career development, how it ensures diversity and fairness, retains talent and so on. This human resource aspect of sustainability is increasingly addressed by major corporations.

– The impact of the products: Products and services have an impact on people. Sustainability implies the goal to foster positive and reduce negative impacts. Product safety is a major aspect for many products and of course quite different, for example, for a car and a can of milk. Other product aspects with a direct impact on people include for instance nutrition scores, harmfulness, product residues, customer data protection and the online security of services. How products and services could create a positive impact on people is a focus of chapter 6.

– Social aspects in the value chain: People across the value chain have become a major sustainability issue as is described several times throughout this book. Addressing people in the value chain includes the minimum of complying with the internationally agreed principles of the International Labour Organization (ILO) on human rights, wages, freedom of association, discrimination, as well as forced and child labor, among others. Auditing suppliers and helping them advance are examples of farther-reaching actions.

– Community involvement: Companies engage in many ways for social causes that benefit people. Examples include corporate donations and sponsoring, volunteering, community projects near factories and so on. It is about giving back and contributing as a company with the resources given to solve societal challenges. (see chapter 8)

The people aspect is becoming increasingly important and linking it to sustainability encompasses different topics and spheres. At this point, I would like to particularly address bigger companies, in the light of the major challenges ahead in the work sphere, to take on an active role. With their power and innovative creativity, companies can significantly contribute to solutions to societal challenges such as unemployment, education, technology and other sub-topics in the future of work mentioned.

7.02 ——————— CALL FOR ACTION

Work and sustainability are both topics that have a significant impact on our future, and at different levels: global, individual and corporate. I am convinced that we have the broad freedom to shape both of these topics, also in places where they intersect. Much has already been said in books, blogs and reports about specific issues such as talent hiring, employee development, smart technology application, the gig economy, and new remote work business models, just to name a few. The people and social spheres are also highly individual to a given company setting. Therefore, I would like to provide my perspective and highlight specifically two aspects of the intersection of work and sustainability that are not too often addressed yet.

CHANGING THE PERSPECTIVE

That companies are citizens with enormous influence and significant resources is one of the general convictions throughout this book. At this point, I would like to focus on the term "shared responsibility" that is increasingly used and, sometimes, also referred to as "shared value" and "social contract". What does it mean?

Multinational Nestlé uses the term "shared responsibility" with respect to employment. One of the company's strategic pillars is people. As a major employer, Nestlé wants to provide a workplace that treats people with dignity and respect. The company currently employs more than 291,000 people around the world (down from 335,000 in 2015) from 150 nationalities. Its products are sold in 193 countries, and it operates more than 400 factories in 85 countries across the globe. They strive for equal opportunity and a living wage. "Providing decent employment and diversity" also comprises one commitment with an outward focus: The "Nestlé needs YOUth" initiative. This initiative, launched in 2017, is striving to help 10 million young people worldwide to access economic opportunities by 2030. Nestlé's motivation reflects an understanding of shared responsibility:

"Employability is a key ingredient of social development, especially employability of younger generations. However, according to the ILO, two out of every five young people are unfortunately either unemployed or have a job that keeps them in poverty. We believe that communities cannot thrive if they fail to offer a future for younger generations. That is why we are determined to help young people develop their skills so that they can find jobs or create their own businesses. This will help build thriving, resilient communities and support the United Nations Sustainable Development Goals. Developing youth helps our business too, because young people are the employees who will keep our company dynamic and competitive, the farmers who will grow the crops we need, and the entrepreneurs who will help us reach new markets."[112]

Nestlé also leads an agripreneurship program that trains and supports young farmers. With less than 5% of farmers worldwide under the age of 35, jobs in agriculture are attracting fewer young people at a time when the world's population is growing rapidly. Nestlé helps to inspire, train and enable the next generation of agripreneurs by providing them the knowledge, skills and entrepreneurial thinking they need to manage farms in the 21st century.

Let's switch now from a company to an industry; namely, to the pharmaceutical industry. Providing for the people who need it, regardless of income, is a major challenge. Two billion people worldwide today live without access to medicine or robust health systems. According to the UN, these people lack access to diagnostics, vaccines and treatments because they are unaffordable, inaccessible or unavailable in their country, especially people living in low- and middle-income countries. Access to medicine includes medicines for neglected tropical diseases, HIV/AIDS, malaria and tuberculosis, as well as child and maternal mortality.

Shared responsibility means that companies take on their position within the challenge: *"The access to medicine challenge is multi-faceted and many different actors must take responsibility. This includes the scientific research community, local governments, public health and regulatory agencies, overseas development agencies, philanthropists, trade administrators, the nonprofit sector including product development partnerships, and both the research-based pharmaceutical companies and manufacturers of generic medicines. Pharmaceutical companies, with the resources and the knowledge to develop medicines, have a responsibility to ensure these technologies are made available to people, regardless of their socioeconomic standing. Improving access to medicine creates new routes to market, opening up demand for new and adapted products. As companies enter new markets, they must do so ethically and responsibly."*[113]

How can the pharmaceutical industry do more to help the world's poorest people access the medicine they need? This question prompted a Dutch entrepreneur to establish the Access to Medicine Foundation in 2003. Since 2008, this foundation has been publishing the Access to Medicine Index every two years. It was the first index to focus on a specific industry sector and topic related to corporate social responsibility. The Index ranks 20 of the world's largest research-based pharmaceutical companies according their policies and practices in place to improve access to medicine, vaccines and diagnostics in 106 low- to middle-income countries, using 69 metrics, and covering 77 diseases, conditions and pathogens.[114]

TIP

The Access to Medicine Index.
✈ tinyurl.com/y6tpv52m

A report assessing the 10-year
progress of the industry
and revealing thrilling insights
into industry dynamics and
corporate responsibility.
✈ tinyurl.com/y3k63nu3

British multinational GlaxoSmith-Kline (GSK) ranks first in the Access to Medicine Index. The focus of GSK's business core competencies is on access to HIV pharmaceuticals and—as the world's largest vaccine company—on vaccination. In 2019, they reached 640,000 people to address the HIV stigma and support HIV eradication through a partnership with Positive Action for Children. GSK also delivered 200 million doses of oral polio vaccine to over 40 million children through UNICEF in support of the Global Polio Eradication Initiative. This is an example of shared responsibility at a corporate level as part of an industry effort and reflects an active contribution to solving societal challenges.
✈ www.gsk.com/en-gb/responsibility

THE DANGEROUS P-WORD, PURPOSE

"Purpose" and "why" are new buzzwords that have entered corporate boardrooms. "How can a successful company embrace purpose" asks con-

TIP

A condensed version (5:01) and
the original (18:02) of "Start with
why" can be found on YouTube.

sultancy Roland Berger. Simon Sinek´s TED talk "Start with why" is among the most viewed TED talks ever. It is an excellent, insightful talk addressing the "why", primarily from a marketing perspective.

There is not yet a common understanding of "purpose" in today's corporate world. This implies a danger of being overused and, consequently, distracting from a good overall idea. Interfaces to sustainability are often vague. With companies wanting to be "good" and perceived as citizens who actively contribute, connecting purpose to sustainability is eagerly pursued. Yet, from my perspective it is another "watchout area". Recently, I recommended to a client that he restrain from embarking on a purpose project as it would not have been credible in the given circumstances where still much homework in the area of sustainability needed to be done.

With that initial warning, let´s look in more detail at what the purpose topic is all about. Initially, it is about carving out the "why" of a company: why does it exist, which problem is it supposed to solve, and how does it want to interact and shape the people it works with. It interferes with our sustainability perspective when it goes beyond the marketing sphere towards the topics of social responsibility and value contribution, as argued for example in chapter 1.

A major driver of the debate on purpose is a new generation that has entered the workplace: The so-called Generation Y or millennials. Born between the early 80s and the late 90s, it follows the baby boomers and Generation X and has a very different mindset than its predecessors. According to several studies, millennials represent a dramatic shift in values combined with self-confidence and high expectations. At the same time, they incarnate ambivalence. This generation grew up in a global world without borders and with multiple options, which was also marked by different crises. They have a high affinity for technology from growing up with the internet and social media. Millennials overall were raised in wealth and security, received attention and appreciation, to the point of being spoiled, leading to high ambition and autonomy. Self-realization is highly valued and having personal priorities and keeping all options open is a typical mindset. In work they are ambitious, but at the same time they are the first generation to rank time over money and purpose above everything. They are curious and flexible but also conservative in several

> **TIP**
>
> An interview with Simon Sinek on millennials giving great insight into the minds of the generation and the implications for the corporate world.
>
> ⟿ tinyurl.com/y6smgfcl

aspects. Their major motivation is to take on a job that provides meaning and gives them a sense of contributing to a higher cause. As soon as they have basic needs and income covered, they strive above all for purpose and personal development rather than status, more money or leadership positions, as in previous generations. They want to be proud to be part of their organization and rank common values extremely high. This overall typology covers a privileged cohort of people, especially the well-educated, living in rich countries and in jobs with high flexibility. Thus, it is not representative for all people or all economic sectors but is a major driver for change, particularly as millennials in 2020 represented about half of the global workforce. It is crucial for companies to respond to these developments. Many companies claim they are pursuing other causes rather than merely making profit. Purpose now seems to be giving companies another boost, a new context, triggered by the employee's desire to belong and be proud of a company.

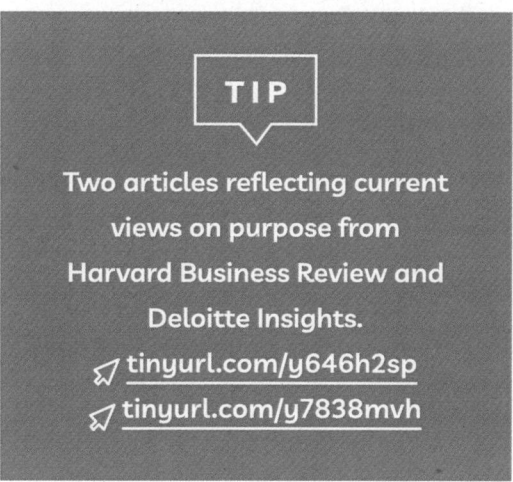

TIP

Two articles reflecting current views on purpose from Harvard Business Review and Deloitte Insights.

tinyurl.com/y646h2sp

tinyurl.com/y7838mvh

As an addition to the initial warning, I would like to emphasize the lack of a clear definition for the concept of purpose. As the understanding is vague, you will encounter a broad range of topics and approaches in conducting your research on "purpose-driven companies". For some this concept is about creating deeper connections with consumers and creating amazing customer experiences. For many it is about attracting and retaining talent. Often, it is about excellence, fluid hierarchies, appreciating employees, and generally creating a workplace that inspires employees, not only physically but also emotionally, intellectually and in aspirations.

Surely there can also be connection to sustainability. Transforming the business beyond mere profit maximization, engaging with stakeholders, providing solutions to societal concerns, serving a bigger cause—many of these aspects are addressed throughout this book. I present several examples of companies that define their purpose by creating value for people and communities. They strive to make a difference in the world through solving a sustainability problem or doing good while doing business well.

That said, I personally struggle with the word and approach of purpose as it is currently practiced predominantly as a marketing approach. I believe that bold statements and self-assurance or yet another management concept are counterproductive. At the core of purpose is credibility and authenticity, which cannot be imparted by campaigns or words.

The posture of purpose is hard to create, implant or transfer and requires true commitment. I found an amazing example of this many years ago in Northern Germany. I had the opportunity to work for several years with Lebensbaum, a mid-size company created 40 years ago by a passionate entrepreneur. The company grew successfully with a great sustainable business approach. It maintained its high values and solid stance amidst challenging market dynamics. The following excerpt describes how specific values have been implanted into the business.

"The Lebensbaum tea, coffee and spices come 100% from ecological farming. We are convinced that this sophisticated form of farming is a prerequisite for high-quality, natural food. But it is about more than this: with every product sold, the number of fields farmed organically increases. We are thus promoting a form of agriculture that understands how to preserve our natural resources while producing foodstuffs of exceptional quality. We buy at the place of origin because quality has roots and we want to accompany the creation of our products from the very start. We see supplier relationships as long-term quality-based partnerships. Together we shape the entire value chain from the field, via the processing in our production facility, all the way to the finished product ready to sell. Business management understood in this way creates lasting value and is beneficial for everyone involved. Putting all modesty aside, we like to call it the food economics of the future."[115]

This is not rhetoric. You can see, hear, taste, and feel Lebensbaum's purpose in their products, their communications, at their company's headquarters, when talking to executives or the front desk employee. They live up to their over 40 years of established, deeply rooted convictions of organic business, continuous improvement and caring for others. Their sustainability report describes in detail their goals, approaches and values. ✐ tinyurl.com/y2qgbcrq

A second example points out the fragility of purpose. The Body Shop started up in 1976, integrated sustainability aspects into its products, its value chain and its interaction with customers and society. It was the first cosmetic brand to prohibit animal testing, introduce Fairtrade, foster supplier communities and stand up against domestic violence. It was a forerunner and role model in the cosmetic industry and for sustainable business–with an early and comprehensive approach. The Body Shop's founder Anita Roddick described her motivation as follows: *"My passionate belief is that business can be fun, it can be conducted with love and a powerful force for good."*[116]

In 2006, The Body Shop was sold to L'Oréal. The experiment of transferring the purpose of The Body Shop to the multinational cosmetics brand failed. In 2017, L'Oréal sold The Body Shop to Natura Group, the largest Brazilian cosmetics multinational. The adventure did test but not destroy the core of the brand. The Body Shop now finds itself again in a family with a shared purpose and a common commitment to sustainable and ethical business practices, as they state on their website. The point is not about "good" and "bad" companies. It is about the fact that purpose does not automatically trickle down and cannot be easily adopted. It is evident that purpose connected to sustainability is very demanding in terms of an honest commitment to its core values and hard work.

7.03 ———— WORKING IN SUSTAINABILITY

Sustainability has become a job sector all its own. The position of sustainability manager, for example, was practically non-existent 15 years ago. With the evolvement of sustainability, especially over the past decade, numerous jobs and work fields emerged. There are no specific categories yet, as many different job profiles could be put under the same umbrella of sustainability. I would like to look at major developments and share some of my observations about the overall issue of "good jobs", which is of particular concern for younger generations like millennials.

SOCIAL OR GREEN?

Social businesses (also called social enterprises) connect the future of work and sustainability in a smart way. Social enterprises have both business goals and social goals, with the main purpose being to promote, encourage, and make social change. The purpose to solve a societal or environmental problem is executed through commercial strategies and a business approach. In comparison to typical businesses, the goal is to be financially stable but not generate a typical profit.

> **TIP**
>
> A 2016 article listing the most influential academic research and describing the gap between the research and practice of social entrepreneurship.
> ⟋ tinyurl.com/y62o9oad
>
> US-based Stanford is an example of a university focusing on social entrepreneurship.
> ⟋ sehub.stanford.edu

For this reason, a surplus from the entrepreneurial activity is usually re-invested into a social cause to maximize social impact. Another goal is often to expand or replicate a successful model in other communities or for other social challenges to generate more impact. Helping social entrepreneurs become sustainable and replicate social business models is an example of one of the goals of social accelerator Ashoka (see page 180).

The social business approach has its roots in Bangladesh and was piloted by Nobel prize winner Muhammed Yunus. It was originally focused on addressing people at the "bottom of the pyramid" (BOP), the poorest people in the world. Providing nutrition, access to healthcare and employment, among others, through business solutions, was the core idea. Today, Yunus multiplies that idea through an accelerator to support social businesses around the world.

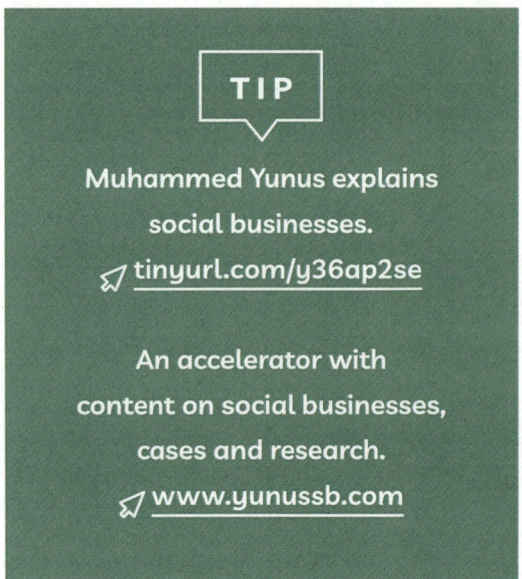

TIP

Muhammed Yunus explains
social businesses.
⤤ tinyurl.com/y36ap2se

An accelerator with
content on social businesses,
cases and research.
⤤ www.yunussb.com

In Germany, for example, "Share" was built as a brand to provide social impact through business. *"We want to facilitate the access to food, drinking and hygiene for all"*—is the mission of this company, which started up just a few years ago. They create "share products", to which direct help in different projects in the three impact areas are connected. ⤤ www.share.eu

In countries across Europe and among young professionals, social businesses are vastly popular, and a growing field for job opportunities. Accelerators and business hubs (for example, Social Impact Lab and the Impact Hub Network) as well as the growing number of crowdfunding platforms such as startnext in Germany facilitate and make it more accessible to develop a project into a social business. Social businesses have been an interesting topic of academic research for several years and increasingly receive a home at universities.

TIP

The video shows how the
European Union supports social
enterprises in Europe.
⤤ youtu.be/f8LUtMZqpHY

"Green jobs" don't have a clear profile, either. While social businesses emerged more from a grassroot level, green jobs were triggered more by politics as a major driver for job creation, for example, through clean technologies. The International Labor Organization (ILO) defines "green jobs" in the following way:

"Green jobs are decent jobs that contribute to preserve or restore the environment, be they in traditional sectors such as manufacturing and construction, or in new, emerging green sectors such as renewable energy and energy efficiency.

Green jobs help:

– Improve energy and raw materials efficiency

– Limit greenhouse gas emissions

– Minimize waste and pollution

– Protect and restore ecosystems

– Support adaptation to the effects of climate change

At the enterprise level, green jobs can produce goods or provide services that benefit the environment, for example green buildings or clean transportation. However, these green outputs (products and services) are not always based on green production processes and technologies. Therefore, green jobs can also be distinguished by their contribution to more environmentally friendly processes. For example, green jobs can reduce water consumption or improve recycling systems. Yet, green jobs defined through production processes do not necessarily produce environmental goods or services."[117]

The ILO has a wide range of ongoing initiatives to promote the growth of green jobs in different countries and industries, develops guidelines and capacity building tools and provides on its website a broad variety of resources on the topic. The aspect of a job being "decent" connects to the SDG 8 "Decent jobs".

A WHOLE NEW JOB WORLD

As mentioned, there are not yet clear definitions or statistics on jobs in sustainability. With a growing importance in different areas of life and business, however, sustainability is naturally turning into a field with many and diverse employment opportunities. This dynamic works two-fold: One, existing jobs integrate sustainability aspects into their job profile. Two, sustainability aspects express themselves in new job profiles and new job opportunities. Let's look at both dynamics.

(1) Existing jobs are integrating sustainability aspects into their profiles. Tasks and requirements related to sustainability are emerging in a wide range of functions, as we see throughout this book. This applies, for

example, to managers and employees working in the areas of health and safety, quality, supply chain management, communication, facility management, human resources, research and development, marketing and so on. Naturally, it also applies to business lines and operations. Many if not all entities of a company have specific tasks to fulfill so that sustainability becomes an integral part of a company's journey (see pages 133f.).

Additionally, in many different industries, demand creates requirements for specific sustainability know-how that needs to be applied to new tasks in the industry. For example, in the building industry supply chain, demand for sustainable construction and eco-efficiency is rising. This in turn requires architects, building material companies, planning bureaus and others to apply sustainability aspects and create new solutions. The same holds true, for example, for value chains in mobility, public infrastructure, agriculture, waste management, education, auditing, public relations and marketing, consulting and so on. These types of basic and specific know-how of sustainability will be needed.

(2) Sustainability aspects express themselves in the form of new job profiles and new job opportunities. Typical examples are the roles of sustainability manager, environmental accountant, sustainability program coordinator, sustainability specialist and climate protection officer. As a result of new initiatives mentioned throughout this book, such as Carbon Disclosure Project, Round Table on Sustainable Palm Oil, Yunus Social Business, Viva con Agua, and Dialogue in the Dark, as well as in institutions such as the United Nations, new jobs are created that deal with implementing the programs and businesses.

Additionally, numerous new job profiles and positions are still arising. The following list includes job positions found in a quick search on different job platforms in the summer of 2020. They are illustrative, and the jobs will often have a different name in a specific context. Here is a random list of selected new sustainability-related job positions in alphabetical order:

Air Quality Forecaster, Biofuel Production Operator, Circular Business Lead, Climate Analytics Consultant, Community Greens Executive Director, Community project manager, Corporate Relations Analyst, Cradle-to-Cradle Scientist, Data Scout Sustainability, Development

Expert for Sustainability Services and Applications, Diversity Specialist, Ecotourism Guide, Environmental Initiatives Program Manager, Environmental Public Relations Specialist, ESG Research Analyst (Controversies and Global Norms), ESG Research Analyst (Weapons and Defense), ESG Specialist Client Relations, Feelgood Manager, Global Empathy Leadership Group Member, Global Marketing Manager Sustainability, Global Strategic Sourcing Manager, Internship in the field of New Venture and Innovation Lab, Junior Strategist Circular Economy, Landscape Architect, Lead Social Compliance Auditor, Life Cycle Assessment Consultant, Project Manager Products Sustainability and Maintenance, Renewable Energy Engineering Internship, Soil Conservation Technician, Solar Engineer, Sustainability Reporting and Outreach Coordinator, Team Lead International Packaging Development, Transformation Manager Global Sustainability, Turf Scientist, User Experience Designer for Sustainability Services and Applications, Water Resource Engineer, Watershed Manager, Watershed Science Technician, Wetland Specialist, Wind Energy Engineer.

From the job positions, it becomes obvious, that specific know-how in the respective area is needed. Besides that, some overall key qualifications are usually needed for sustainability-related jobs. They include, among others, interdisciplinarity, transdisciplinary thinking, ability to learn, cooperative abilities, team and communication skills.

WHAT IS A "GOOD JOB"?

I found that many people today—both young and those in the second phase of their careers—search for a "good" or a "meaningful" job. Especially over the past 15 years, I have spoken with hundreds of people in interviews, people who reached out to me, as well as those in my professional and church networks. Many of them wanted a job in sustainability, at an NGO or social startup, in sustainability consulting or just in "something good". I have met many people who have thought of quitting their "regular jobs" to switch into "good jobs". I believe this is one of the most underestimated job transformations taking place in countries like Germany. We need to respond to it.

It became a habit for me to ask people: Why are you searching for a good job? And, later in the conversation, I would follow up with: What is a good job, and how do you define meaningful?

They always turned into great conversations about passion, deception, expectations, and world views. And, despite a variety of stories, one motive stood out clear: People search for purpose. We have already heard that several times but let me approach it from a personal perspective that has evolved from these many conversations.

It is evident that people want to make a difference, change things for the better, work in a job they could be proud of, contribute to something bigger and beyond themselves. Most of these people—young and older— are motivated by a deeper desire of belonging to something bigger. Many basically expect increased value and self-worth for themselves from such a job. And ethically speaking, they wanted to be on "the right side, the good side". This is a noble and understandable motivation.

When people ask for my advice, I give them a very personalized answer but still usually try to make clear two points I learned through all of my conversations over the years, as well as from my own personal journey on this topic.

First: It's not the task of a job to make you happy. It is crucial to understand this. Otherwise our work lives are likely to become miserable. No job can provide what you really long for, not even the best job in sustainability. Because what you long for, the desire underneath the desire, goes deeper than your job. Economically speaking, we have the wrong allocation of resources, because people are looking for the right things in the wrong places at the cost of the common good. Spiritually speaking, the hunger for belonging is nurtured in the wrong place.

Second: Whenever possible, I encourage people: Try to be the change you want to see in the place where you are in. Why? Because we need transformation in all spheres of society. If all motivated people go to the (supposedly) "good jobs", society won't change sufficiently for the good. We need startups but also change in big corporations with huge impacts– that is where the game will basically be decided, as pointed out several times.

And there is another, often underestimated aspect: the fact that it can be helpful to already acquire the skills now that will ultimately be crucial for change.

Let me illustrate this with the example of a young woman who over the years became a friend. She first reached out to me about five years ago because she wanted to switch to a sustainability job. She had been working for a couple of years in consumer goods marketing, and quite successfully, but did not really enjoy it—or, to state it more precisely: did not see the value in her job. We had long conversations about her motivation, the value of a marketing job, and the not-so-perfect world of sustainability. After that, she asked for advice and wasn't very happy with the answer: For the moment, keep your job, learn as much as you can about business, numbers, marketing and campaigns, and learn to love the place you are at now. In the years that followed, she was promoted several times, switched to new positions and acquired knowledge, experience, recognition and a network. At some point, out of the blue, a sustainability project came up and she willingly invested the extra time required for this in addition to her regular job. About six months ago, she was offered the job of sustainability manager at the European headquarters of her company. Among the major tasks was aligning the different product segments to sustainability, coordinating value chain projects for products and driving sustainable packaging. Because she had worked for many years in all different product segments and positions, and on operational and strategic projects, she knew all of the different perspectives—company, customer, stakeholder—and had a distinguished network. At that point, she was more than prepared to serve her company with all of her added skills, much more than any other person, including herself five years earlier. Needless to say, she accepted.

APPLICATION

1. ———— Reflect on the following questions:
   – Which jobs do you like and think are valuable?
   – Which jobs do you think could be eliminated?
   – Which thoughts, assumptions, and experiences
   have shaped your assessment?

2. ———— The simple survey about what you do at work will
   reveal how your job will likely change due to automation. ⤴ www.oecd-futureofjobs.org/start

3. ———— Read some stories about people whose working lives
   are probably less fortunate than yours.
   ⤴ www.kiva.org/about/impact/success-stories

4. ——— You may also want to find inspiration in learning
about how little is necessary to kickstart decent work
opportunities across the world by investing in people.
⊿ www.kiva.org/lend-by-category

5. ——— If you are searching for purpose or a "good job", ask
yourself the following questions. You can also ask a
good friend that may be in the same situation how
they would answer these questions.
– What do I define as "good"?
– Why? What expectations, beliefs,
motivations do I relate to?
– What are the underlying needs and desires that not
even the best job in the world could fulfill?

CONTRI-BUTING TO SOCIETY

In early 2000, McKinsey, my employer at the time, launched an initiative in Germany called "startsocial". It offered projects in the social sector the opportunity to apply for free advice from businesspeople on strategic and operational issues. Back then, providing resources to do good was a pretty novel idea. Meanwhile this is widely practiced and has developed into a relevant contribution.

It was fascinating to see how much value could be contributed with the addition of some business know-how. And the reverse was also true. To be able to enter the new world of worthy causes was powerful. Many businesspeople did not interface much with people marginalized in society or struggling to make a living. The passion people in the social sector shared for their cause and the creativity necessary to work effectively with limited resources impacted even the toughest business folks. Both sides met at eye level, and society became a bit more permeable. These experiences have shaped me fundamentally.

——————— # WHAT BUSINESS CAN DO FOR SOCIETY

The major focus of this chapter is the outer circle of the scope of influence—that is, the activities beyond the core business and the value chain, targeted at society. It is the "value-add" companies bestow on societies, their voluntary contributions in the sphere of society. Obviously, spheres often intersect. It is not uncommon to see activities in the outer circle interfere with activities in the core business or in the value chain, as we will see.

WHY COMMUNITY ENGAGEMENT HAS SURGED

Peter Drucker (1909–2005), an influential thinker and professor on management theory and practice, followed a simple thought regarding corporate responsibility which can be phrased as: *"It's about what a business does TO society, and what it can do FOR society."*[118]

Much of this book focuses on what business does TO society, how to mitigate the negative impacts, and in what way can the social and environmental demands directed at a company be actively managed. This chapter, in contrast, focuses exclusively on what companies can do FOR society.

Companies contribute to society on a voluntary basis as citizens in the communities they operate in. This is why one of the terms commonly applied is corporate citizenship. There are several other terms, including community engagement, social engagement, community investment and giving back. Philanthropy, corporate giving and donations are further terms applied in this area. Though their understanding differs a bit from a conceptual point of view, some of these terms are frequently used interchangeably.

A major area of involvement for companies is communities. The reason is the multiple interfaces companies have with people and nature on a local and regional level. Local community engagement is strong, especially in countries and regions where companies have their factories or headquarters located, or where they source their raw materials or sell products. This is why I am devoting a special passage to this topic.

Traditionally, corporate contributions to society are significant as well as diverse. Since the 2000s, engagement has increased and more strategic approaches have emerged. This was coupled with a new dynamic in the social sector—the mushrooming and spreading multiple of new initiatives, the professionalization of organizations, and the establishment of social venture funds and other supporting structures. Demand for financing social causes rose and nonprofit fundraising in the corporate world came in vogue. Several developments led to transparency and accountability becoming more important in the social sector.

Transparency increased, the governance of contributions was formalized and more strategically brought in line with the overall sustainability activities, especially in larger companies. Structures and processes for corporate giving became more systematic, for example, through guidelines, a clearer focus and reporting. Strategic approaches emerged.

WHAT IS STRATEGIC SOCIAL ENGAGEMENT?

Social engagement occurs in the outer circle of a businesses´ sphere of influence. It encompasses the voluntary positive impact of the company and its contribution to the world's needs. It should be emphasized that "social" can refer to aspects related to people as well as to nature and, as such, is contributing to society. Obviously, the social engagement of a company is highly individual. This can be attributed to fact that companies differ in their resources, goals, preferences, places, history and status quo of corporate social engagement.

Should companies contribute just "anything"? This would be not the best approach. By their nature, companies act from an economic standpoint. They allocate limited resources to produce a certain output that serves needs and, ideally, creates a value-add. This logic increasingly extends to engagement in society as it also has limited resources—money, time, and skills— to spend. As significant resources are allocated, awareness and exposure expand, and the need for strategic alignment rises.

Companies will often define the goals they want to achieve through their engagement. Some objectives for social engagement may include one or more of the following:

– Helping communities related to the company flourish

– Establishing a positive reputation with the public

– Regaining or strengthening the social license to operate

– Motivating and attracting employees

– Contributing to a greater social cause using company resources

– Alleviating a need in the world

– Supporting business goals such as consumer bonding and a secure supply base

– Broadening and deepening employee skills, such as social competence

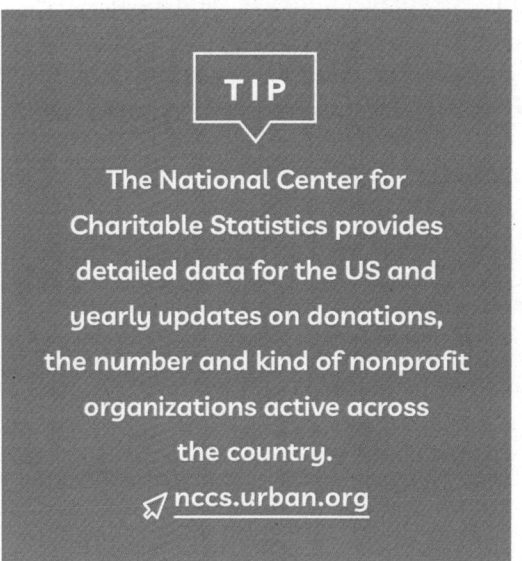

TIP

The National Center for Charitable Statistics provides detailed data for the US and yearly updates on donations, the number and kind of nonprofit organizations active across the country.

↗ nccs.urban.org

While it is useful to set objectives for social engagement, after many years of experience, I find it even more relevant to carefully consider the resources the company applies and to choose an approach that fits not only to the specific need but also to the company. This will depend on the credibility of the engagement and the impact created for the cause. Strategic alignment can have two directions: Either there is a clear idea as to what resources the company wants to apply—for example, a defined amount of money or certain skills to address a specific issue or cause. From that point, it is easy to derive and refine an approach. Alternatively, an existing portfolio of donations and projects could be better targeted for more impact or better aligned to the company's sustainability agenda. Strategic engagement thus means that companies focus their efforts on specific topics, activities, and programs and often partnerships.

From a status quo assessment, a plan for resources and an approach can be developed.

For successful strategic engagement, I suggest the following four criteria:

– It should be intentional rather than reactive or random.

– It should be focused, which can mean either solving a social challenge, supporting a specific target group, setting a regional scope or similar.

– It should leverage the company's specific resources.

– It should strive to create a significant impact.

There is an additional, fifth criterion to ensure well-structured, strategic social engagement, covering a broader context, which will be assessed at the end of the chapter.

THE ABUNDANCY OF RESOURCES

To build an intentional, focused program that creates impact and leverages company specific resources, it is useful to assess which resources can be applied. Companies possess numerous resources that can add value to people and nature. Over the past two decades, an impressive variety of social engagement undertakings has emerged—and we will see more of this. Business skills and competencies, in particular, have increasingly emerged as part of social engagement approaches and have certainly outpaced traditional money giving.

Let's look at some typical resources that companies possess and which of these they could possibly provide to social causes.

Please note that the resources depicted in the table are illustrative and different resources are often combined. For example, in the mentioned Google.org Impact Challenge on Climate, the company mainly provides a significant amount of money. However, the selected initiatives will receive also support knowledge from Google's business accelerators—thus, beyond money, the company provides its skills and an existing infrastructure (its accelerator). In the Disaster Help Teams, DPDHL contributes its network but also engages through time donated by employees who volunteer.

The table below gives an initial overview of some of the typical corporate resources and how they can be applied to social projects. A "perfect match" is highly individual.

Corporate resource	Applied to a social cause	Examples
Money	Donations, sponsoring, foundation endowments	See the examples in disaster relief and corporate foundations. The Google.org Impact Challenge on Climate commits 10 million euros to fund ideas that use technology to accelerate Europe's progress toward a greener, more resilient future.
Products or services	Donations of products, provision of in-kind services	Apple donates products to schools. Kellogg's donates cereals and bars to food banks across the world. Several law firms advise individuals and organizations with social causes pro bono on legal issues.
Time	Volunteering of employees	startsocial unites employees of German companies to donate time and know-how to community projects and social initiatives. PIMCO, a global investment management firm, engages every year in a month of volunteering to demonstrate its commitment to communities. During this "Global Month", employees, friends and families volunteer in local initiatives.
Retail capacities	Provision of logistics for marketing and distribution of social causes	Wall, Germany's biggest supplier of City Light Posters, donates advertising space in cities to selected social causes and initiatives at no cost. Through smile.amazon.com, Amazon donates 0.5% of the eligible purchases to the customer's favorite charitable organization.
Networks and infrastructure	Providing contacts, corporate structures (e.g., logistics, value chains) for a social initiative or cause	The Disaster Response Teams (DRTs) are part of DPDHL Group's GoHelp disaster management program, which it has operated in partnership with the United Nations since 2005. Through this partnership, the DPDHL Group provides the UN and country-level disaster management agencies with pro bono access to its core logistics expertise, and the logistics skills of more than 500 specially-trained employee volunteers worldwide who can deploy within 72 hours after a disaster.

Corporate resource	Applied to a social cause	Examples
Skills and competencies	Know-how, ideas, business skills etc., are targeted to solve a social problem	Siemens Smart Clinics provide access on a charity basis to primary healthcare, which combines basic medical equipment with a secure energy supply. Siemens Smart Clinics are in operation in Jordan, Iraq and Columbia and planned for Egypt. US-based provider Waste Management conducts an annual Waste Management Phoenix Open Tournament, a huge zero waste sports event to educate, raise awareness and promote recycling.
Reputation	Marketing and advocacy for a social cause, cause-related marketing	Airbnb stands up for equal rights and diversity with a campaign called "We accept", featuring a short ad: ⟿ youtu.be/yetFk7QoSck Beer brand Stella Artois runs a longstanding cooperation with Water.org to raise awareness and provide access to water to women and their families in the developing world. ⟿ youtu.be/mHnVMfjFMVs

Own research and structure

The table lists a range of resources, whereby the first two categories—money and company products and services—are relatively easy to provide. When giving them away, the company has little executional control over the handling and impact of its giving and is relatively little involved. That's why companies are increasingly looking to apply resources, such as business skills, either instead of or in addition to donating money. As you move down the resources in the table, the complexity of setting up a project rises—and even more so when combining resources. That's why it becomes important to have a strategic and systematic approach in order to find the right setting and manage it effectively. This also helps to achieve the impact for the social cause—the most relevant objective within social engagement.

SUPPORTING LOCAL COMMUNITIES

Let's zoom to the important area of investing into local communities. Creating value for society, specifically in community surroundings, often includes giving to local communities and investing in employment. We will also look at disaster relief—an area where companies significantly contribute, often with little recognition or appreciation from the public.

GIVING MONEY—A LOT

Many companies across the world donate money to local communities. They have often been contributing for decades to the wellbeing of their neighborhoods without saying much about it. Still today, many companies, especially smaller ones or family owned businesses, donate without many administrative hurdles and without much publicity.

Money is given to all kind of causes and target groups, for example, to people and initiatives that support children, elderly people, the poor, homeless, refugees, disabled or otherwise disadvantaged people, to local youth initiatives, arts, sports, environmental protection, improvement of neighborhoods and alike. Besides money, other resources are also often provided, for example, employee time or company products and services, when appropriate. These contributions are usually called in-kind donations.

The charitable giving statistics for the United States show that companies donated 20.77 billion US dollars in 2017, an 8% increase from 2016. Walmart was the leader, donating 311.6 million US dollars in cash in 2017.[119]

A typical motivation to donate is described well by the statement of Honda: *"We may be a global company but, at our core, we are people—people who care about our families, our neighbors and our communities. Our programs and our charitable giving are how we are trying to make a difference in our neighborhoods and around the world."*[120]

The automotive company gives *"time, talent and resources to enrich and strengthen the communities where we do business and where our associa-*

tes live and work". Two specific applications: One major topic of Honda's engagement is youth education, especially in STEM (Science, Technology, Engineering and Math) with several programs, as well as grants and charitable donations. Another project intersects more directly with the business as it involves customers: Through the program "Ride for Kids", motorcyclists in the US through the Honda Riders Club have raised 71 million US dollars over the past several years for kids with brain tumors.[120]

In the case of CEWE, Europe's leading photo service provider, sustainability encompasses social commitment and is therefore one of CEWE's five sustainability pillars. Local communities around the factories located across Europe, particularly benefit from funding for social causes, education, photo culture and environmental initiatives. In 2019, 1 million euros were granted for social initiatives. A significant amount went to SOS Children's Villages across the world, several of them cooperating at a local level with CEWE's plants. Another example included a social photo project in Hungary: A photographer initiated a campaign to take beautiful pictures of people with Down Syndrome—over 200 photographers eventually participated. CEWE financed the display of the photos and supported the artists as well as the social cause.[121]

Often, donation programs are very much targeted at certain areas or groups. In 2019, Disney gave 338.2 million dollars in cash and in-kind donations *"to nonprofit organizations that bring comfort, inspiration, and opportunity to kids, families, and communities around the world"*. Supporting children's hospitals and granting wishes to deliver comfort and inspiration to children facing serious illness is one major area. Another focus area of Disney's engagement is providing inspiration and opportunity to young people—the next generation of innovators and storytellers.[122]

Worldwide operating Dow Chemical focuses on a specific target group but also regionally with its program called "Dow Promise". The employee-led initiative, started in 2000, strives to positively impact educational and economic challenges faced by youth and adults of African heritage in communities near Dow sites. One pillar is an annual grant program designed *"to support social, economic, or environmental projects that contribute to long-term success in the communities in which Dow operates and in which Dow employees reside ... Dow values the opportunity to provide financial sup-*

port to communities in which the Company has a presence. With great demand for contributions, Dow wants to target its funding in the areas that will have the greatest long-term, positive impact for community members."[123]

Australian based BHP, one of the biggest multinational metal and mining companies in the world, wants to develop strong, mutually beneficial relationships with communities, regions and countries where they do business and contribute to their economic and social development. Community development is targeted *"to enhance our reputation and social license to operate."*[124]

Overall, BHP has made a commitment to invest 1% of pre-tax profits in programs that aim improve the quality of life for people around the world. The 1% is a share that is often referred to a benchmark for the amount of investment in communities, whereas most corporations give less.

BHP invests to support the local communities in which it operates and donates to the BHP Foundation, which invests in global and national projects. In the 2019 financial year, this investment totaled 93.5 million US dollars. Through the Benefiting MyCommunity Program, BHP channeled in 2018 and 2019 community grants of up to 10,000 US dollars to 106 community-driven initiatives across Australia. Among them were a technology project at the Queensland Museum, "Books in Homes Australia" which provides books-of-choice to children living in remote, disadvantaged and low socioeconomic circumstances and a social enterprise that creates work opportunities for homeless, marginalized and disadvantaged people.[125]

Funds for local giving, such as at BHP, are now established by many bigger corporations. With that, transparency increases. Clear processes and guidelines give direction, not only to the corporate grant giver and organizations applying for a grant but also to the public.

The local community grant program of retailer Walmart, for example, gives grants ranging from a minimum of 250 US dollars to a maximum of 5,000 US dollars. Eligible nonprofit organizations must operate on the local level or be an affiliate or chapter of a larger organization that operates locally. They must also directly benefit the service area of the facility from which they are requesting funding. ⌁ tinyurl.com/y3ueawln

The Disney Conservation Fund is an example of a cause-focused fund. It runs additionally to the social engagement mentioned above. The Conservation Fund is *"committed to saving wildlife and building a global community inspired to protect the magic of nature together".* Since 1995, the Disney Conservation Fund has directed more than 100 million US dollars to support nonprofit organizations working with communities to save wildlife, inspire action and protect the planet. ⤢ tinyurl.com/yabeu4d6

SPURRING LOCAL EMPLOYMENT

Employment is part of the inner circle of businesses´ sphere of influence—as it relates to the core business. But employment is also in many areas of the world an important and sensitive topic to societies. Generating and fostering employment in regions where this is a vital and much needed contribution adds a social surplus to communities. Granting support to gain qualifications or facilitating job entry is a major challenge of many societies and will continue to be in the future. Companies as a "natural" partner in this challenge can contribute value in diverse ways, as we saw with Nestlé. (see page 216)

Let's look into three approaches how companies contribute to local employment and for especially vulnerable target groups: employing local staff, investing into employability and integrating refugees into the labor market.

Local employment is valuable in rural areas across the world, also across Europe. As for example sugar producer Nordzucker describes: *"Nordzucker is a European company, based primarily in the rural countryside. We aim to give all our colleagues a rewarding, healthy working environment, where everyone has the opportunity to develop and realize potential. We are also an important partner in local communities, providing employment opportunities and income. In the rural areas where most of our operations are based, our plants are major employers. They offer qualified and sustainable occupations and high-quality apprenticeships, primarily in a technical environment."* [126] The importance of local employment certainly holds true for many developing or emerging countries. ExxonMobil, one of the world's largest publicly traded energy providers and chemical manufacturers, is one of

CORPORATE FOUNDATIONS

Numerous companies have established independent foundations.

In the United States, many advantages come along with forming a foundation, among them tax exemptions and reduced costs. Thus, a significant amount of charity activities in the US are covered through corporate foundations. Companies can build endowments in years when company profits are high that can be tapped during less profitable years.[127]

Across Europe, there is no unitary handling of foundations. In some countries, among them Germany, Finland and Italy, a dedicated purpose has to be established. In Germany, foundations usually are meant to last for eternity and can generally not be dissolved. A foundation only invests the surplus resulting from the assets donated to it, which form the foundation's capital stock.

Generally speaking, foundation models differ and cannot be easily compared. Many different approaches and causes exist, some of them are more or less related to the company's business. Here below is a small selection:

The LEGO Foundation was founded in 1986 and aims to build a future in which learning through play empowers children to become creative, engaged, lifelong learners. Through its programs, millions of children aged 0-12 are reached with learning through play in different parts of the globe. One focus is on play at early childhood, especially for disadvantaged children. The Foundation's Center for Creativity, Play and Learning also conducts research.

www.legofoundation.com/en

The Walt Disney Company Foundation has awarded college scholarships for more than 50 years. 150 academic awards are granted to graduating high school students of Disney employees and cast members from around the world each year. Scholarships are awarded based on students' academic achievement, extracurricular activities, community service and demonstrated leadership. ⚐ tinyurl.com/ybh779w5

The Shell Foundation works relatively close to the company's business scope. Founded by Shell in 2000, it is a UK-registered charity working independently from the company. The Shell Foundation supports pioneering social enterprises and institutions that serve low-income communities across Africa and Asia lacking access to affordable energy and transport. ⚐ services. shellfoundation.org

The Volkswagen Foundation is the largest private research funder and one of the major foundations in Germany. Foundation capital amounts to 3.5 billion euros. Since 1962, the Foundation has granted more than 5.3 billion euros of funding for over 33,000 research projects in a vast range of disciplines. ⚐ tinyurl.com/yxoya6fw

The Honda Marine Science Foundation helps to restore marine ecosystems and facilitate climate change resilience. It supports efforts that improve and preserve coastal areas for future generations and awards grants to respective projects. Inspired by the Japanese concept of sato-umi—the convergence of land and sea where human and marine life harmoniously coexist—the foundation was formed in 2017. ⚐ tinyurl.com/y2xhkpny

Electrolux is a major Swedish company in the appliance industry. The **Electrolux Food Foundation** is an independent, nonprofit organization founded in 2016 and funded by Electrolux. It supports Electrolux employees' initiatives to inspire more sustainable food choices among consumers and professionals, and to support people in need in nearby communities. ⚐ tinyurl.com/yxf4aewd the many examples that adds economic value to countries it operates in by employing and training the local workforce and supporting local suppliers. ExxonMobil develops a local content plan specific to each country or area to establish long-term economic benefits. In 2018, the share of local hiring was across different countries significant; unfortunately, the company does not display absolute numbers or locations with lower ratios to allow a full picture.

	Percentage of personnel locally hired	Personnel locally hired in supervisory or managerial positions
Angola	91%	77%
Chad	92%	80%
Equatorial Guinea	77%	30%
Indonesia	97%	93%
Malaysia	98%	90%
Nigeria	94%	88%
Papua New Guinea	68%	19%

Own presentation based on source: ExxonMobil. 2018 Sustainability Report Highlights. https://tinyurl.com/y7tlv53o. Accessed December 15, 2020.

Directly employing a significant amount of people in developing countries is set as one of three factors for having a positive impact in communities by the World Benchmarking Alliance (WBA). It strives to assess the impact huge, multinational companies have in developing countries.

Employment is described as vital for these countries and has enormous effects on positive impact cycles regarding sustainability. For the impact assessment, WBA sets a threshold when companies directly employ 25,000 or more people in developing countries. This applies, for example, to US-based Walmart with over 280,000 employees across Latin America and sub-Saharan Africa, Falabella (Chile), another retailer, employing over

TIP

A WBA Report entitled "The impact of multinationals in developing countries" highlights the connections of global value chains and the impact of big corporations. It drafts a framework to assess impact in the light of the SDGs.

✈ tinyurl.com/yxmc7vb5

78,000 people in six developing countries, the Swedish telecommunications company Ericsson, which employs over 34,000 people in 11 developing countries and US-based Ford Motor Company, employing over 81,000 people in seven developing countries.[128]

A second approach is employability: to create a workforce that possesses the skills needed in the labor markets. It is now and will increasingly become a major topic in many countries across the world. In a country like India with a large percentage of young people, it is crucial to ensure employability. A promising approach is taken by Tata—an Indian conglomerate combining several different companies to form one of the biggest employers in the country.

"Two-thirds of India's 1.2 billion population is under 35 years … a significant percentage of this population is unskilled or under-skilled. Skill building is therefore a key focus area for national development. … 13 million enter workforce each year. … As per the 11th Five Year Plan published by the Indian Government, there will be a shortage of 55 million skilled workers in 2022 in the rest of the world, whereas India will have a surplus of 47 million skilled workers."[129]

TIP

Training people to become facilitators and to be able to coach people is a way of empowerment pursued by Tata Strive. A short video gives insight into their work.

✈ tinyurl.com/y3jarvq4

With the STRIVE program, Tata's mission is *"to build capacity to train youth for employment, entrepreneurship and community enterprise"*. The skill development initiative reaches to communities, develops skills of people from financially challenged backgrounds and acclimatizes them with the changing work environment. It wants to develop courses that train and provide skilled manpower across the entire industrial spectrum and to foster entrepreneurial talent.

Thirdly, the integration of refugees into labor markets will be one of the major tasks of the next decades, especially across Europe. Multiple and diverse approaches will be necessary and are beginning to emerge.

Scandinavian retailer IKEA announced in December 2019 its *"biggest step in IKEA history taken to support integration of refugees by 2022"*. It plans to support 2,500 refugees through job training and language skills initiatives in 300 stores and units in 30 countries. Furthermore, it strives to create sustainable livelihood for 400 women through a partnership with Jordan River Foundation. The IKEA Foundation commits 100 million euros in grants over the next five years for programs that help refugees and their hosting communities improve their incomes and become more self-reliant. This is another example where business motives—gaining new staff—interfere with a social cause—creating opportunities for people with difficult starting conditions. IKEA believes that refugees with their skills and experiences bring value to business and society, as an executive stated: *"Work is a key driver for integration into society. With this initiative we play an important role in equipping and qualifying refugees to get a job and build their own connections in the local communities. Collaborating with refugees brings new skills, diversity and different perspectives to our business."*[130]

HELPING IN DISASTER RELIEF

Many companies are highly engaged when global catastrophes occur. US corporations for example donated millions of dollars and supplies for the Tsunami that hit Asian countries in 2004. Pharmaceutical and healthcare products companies, as well as other multinationals, were among the largest givers, donating mostly to the Red Cross.[131]

Covid-19 was and is also covered, to a large extent, through the financial support of corporations. This reflects upon an understanding of taking part as citizens and contributing to our society, as exemplarily stated by Jean-Paul Agon, Chairman and CEO of L'Oréal: *"In this unprecedented crisis, it is our responsibility to contribute to the collective effort in every way possible. Through these actions, L'Oréal expresses our recognition, our support and our solidarity towards those who are demonstrating extraordinary courage and selflessness in their efforts to combat this pandemic."*[132]

Seven examples that are representative for the numerous others will be mentioned. They show a broad panorama of corporate resources and focus, as well as a high level of generosity.

– L'Oréal USA donated personal care products valued at more than 1 million US dollars to the organization Feed the Children. Additionally, it announced in early April that it would freeze payments due from very small and small-sized enterprises in its distribution network, including hair salons, until their businesses resume. Furthermore, L'Oréal USA shortened its payment times for small suppliers who have been most exposed to this economic crisis. In mid-March 2020, L'Oréal also launched a Europe-wide coronavirus solidarity program.[133]

– Mars provided an initial cash and in-kind contribution of 20 million US dollars for communities *"where we live, work, source and operate"*. This included 5 million US dollars for the international longstanding partner CARE in order to provide critical supplies and support to women, children and refugees across West Africa, Southeast Asia and other regions. 2 million US dollars went to the United Nations World Food Programme for emergency food and lifesaving protective gear for UN agencies. 1 million US dollars was donated to the Humane Society International to help cats and dogs that have been abandoned, left behind or surrendered, as Mars´ business includes a worldwide operating pet care branch.[134]

– Goldman Sachs deployed over 38 million US dollars in philanthropic support to 27 countries until June 2020. The firm launched the Goldman Sachs Covid-19 Relief Fund with initially 25 million US dollars and committed 5 million US dollars to match employee donations to nonprofit organizations.[135]

– Brazilian-based JBS is the world's largest meat producing company with 130,000 employees in Brazil and more than 240,000 worldwide.[136] As the country was hit extremely hard by Covid-19, JBS launched a program "Fazer o Bem Faz Bem" (translating to: "Do the good, well"), under which it donated 700 million Brazilian real (about 131 million US dollars) to combat Covid-19. A total of 400 million Brazilian real (about 75 million US dollars) were spent in Brazil and allocated to three fronts: healthcare, social action, and science, channeled to 280 municipalities to benefit 76 million Brazilians until September 2020. 300 million Brazilian real (about 56 million US dollars) went overseas, with the majority allocated to the United States where JBS has 60,000 team members in over 50 cities. The priority areas in the US are healthcare, social assistance, and infrastructure.[137]

– Deutsche Post DHL (DPDHL) Group, the world's leading postal and logistics services provider, collaborates on emergency preparedness and disaster response efforts in times of need. The company provided volunteer humanitarian logistics support for the Covid-19 response in Argentina, Brazil, Colombia, Costa Rica, Ecuador, Guatemala, Panama and the United States. The efforts of DHL's employees fall under the company's GoHelp volunteer initiatives and are led in cooperation with the Red Cross.[138]

– Hamburg-based tobacco company Reemtsma has been supporting two local homeless relief organizations for a number of years. At least 2,000 people are officially homeless in the city and were hit especially hard by the lockdown, as help providers were also forced to shut down. Reemstma immediately provided 300,000 euros through its partners to give homeless people access to safe, hygienic shelter in local hotels. Employees raised 25,000 euros for this initiative, which was then matched in the double amount by the company.[139]

– Microsoft committed to helping 25 million people acquire new digital skills needed for the Covid-19 economy. Job seekers can access tools and resources to create in-demand technology skills for free across LinkedIn, Microsoft Learn and GitHub. The US community skills program donates cash grants and capacity building to up to 50 nonprofits that offer digital skills and workforce development in Black and African American communities.[140]

———————— # FIVE TRENDS IN ENGAGEMENT

We saw how diverse resources can be translated into ways of engaging. Additionally, a company needs to decide whether it wants to conduct a singular project or a wide-spanning program, work with partners and with whom, and if it would rather focus on one single topic or spread its focus over several areas of engagement. Each company needs to define its own approach. The following five trends depict some insights and examples of how to strategically develop and align social engagement.

FOCUS—NARROW OR BROAD

Strategic engagement moves companies increasingly away from responding to donation requests and spreading small amounts of money to many different causes. Focus will often mean operating in a few selected areas, engaging directly and partnering.

The airline Emirates is a representative example that states it accordingly: *"Although we occasionally fund a program or organization directly, we more often work with partners such as community institutions to strategically target an impact area."*[141]

The same applies to airline KLM. *"The budget that KLM makes available to sponsorship is entirely spent on the above initiatives. We are therefore, regrettably, unable to accommodate any individual sponsorship requests. ... KLM chooses to develop long-term partnerships, which ensure a clear focus and optimize the efficient deployment of available funding."* [142] KLM focuses on its program "Wings of Support", which runs projects that provide shelter, education and medical care to underprivileged children worldwide. It was initiated by the airline's employees who in the course of their work frequently came into contact across the world with people who were suffering.

Focus can also mean being widespread, if it fits strategically. Walmart pursues this approach, as does the Swiss retailer Migros, who gives financial support to a broad range of activities and projects across the country, including sponsoring events, family and children activities, festivals etc. Additionally, through the "Kulturprozent" (Migros Culture Percentage), it dedicates 1% of its turnover every year to support artists and initiatives in communities, as well as cultural diversity in the country. Through its institutions, projects and activities, it gives the general public broad access to cultural and social events. In 2019, Migros invested 117.9 million Swiss francs through this program. ⊿ tinyurl.com/y4k9z9ry

The strategic aspect is that Migros, through its very broad-based activities, is present in its core market as well as close to the day-to-day activities of its customers in their regions.

Focus for multi-site companies often means setting up a few major focal points and bigger lighthouse projects at a corporate level. It also usually means giving the individual company sites the freedom to define the engagements that fit the need of the local community, as in the examples of Dow and BHP.

GOING FOR IMPACT

Social engagement is moving away from singular projects to bigger, multi-year commitments through specifically focused programs. Connecting business resources to society needs within a specific geographical scope and for a specific target group often generates impact for both sides. Programs have a longer-term perspective and invest a significant amount of resources, ideally in an efficient way. The company dedicates itself to a cause, which often involves partners. Closer interaction through a program-approach also creates a sense of belonging and builds relationships and trust.

Two companies with programs striving for a specific impact are Apple and FrieslandCampina.

Through its ConnectED initiative, Apple donates products and provides support to schools. What makes it a strategic engagement program is the overall setup. Apple donates a significant amount of money by pledging 100 million US dollars over several years. It focuses the resources on a specific target group of 114 underserved schools across the US, providing them with teaching and learning solutions. Apple donates products, such as an iPad to every student, a Mac and iPad to every teacher, and an Apple TV to every classroom. Beyond donations, this program joins people on the pathway by providing planning, professional learning, and ongoing guidance—with the objective that every school should experience the transformational power of technology. Additionally, Apple monitors impact through the independent assessment of the initiative and regularly reports on it. ⚡ www.apple.com/connectED   ⚡ tinyurl.com/y2cjhux8

Dutch dairy company FrieslandCampina has established the Dairy Development Program, a long-term program that intersects between business and societal needs. FrieslandCampina shares its business knowledge to support small farmers in Indonesia, Thailand, Vietnam, Malaysia, China, Romania, Nigeria and Pakistan. Farmers are provided support to improve milk quality, increase productivity and establish environmentally sound practices. Building up vital dairy sectors in the developing world contributes just as much to the economies as to sound nutrition. The company has built long-term relationships and gained a reputation that facilitates its access to future markets. This program has helped 250,000 farmers so far. ⚡ tinyurl.com/y7a3rgyo

ALIGNMENT WITH THE SDGS

The Sustainable Development Goals (SDGs) are a framework to advance sustainable development across the world until 2030. Many companies have adopted the SDGs as a strategic framework for their sustainability activities (see page 111). They also serve to increase the focus on the social engagement by allowing companies to select one or more of the 17 SDGs to define their contribution to societies and to sustainable development.

An example of a company applying the SDGs to social engagement is chemical company BASF. The company connects its overall sustainability activities to the SDGs and contributes to all 17 SDGs across its business operations and value chain. On the BASF website, different action points and programs are allocated along the 17 SDGs.
⚡ tinyurl.com/y3gxjkvv

In the area of corporate citizenship, education is a cornerstone and is thus connected to SDG 4 "Quality education". The SDG 4 specifically strives to ensure inclusive and equitable quality education and promote lifelong learning opportunities for all. As a leading chemical company, BASF aims to stimulate an interest in science and foster curiosity. This is the focus of its education programs. In hands-on chemistry workshops called Kids' & Teens' Labs, girls and boys get to be researchers and discover the world of science. Over a million kids and teens in more than 40 countries around the world have taken part in experimentation programs since the first hands-on workshops in 1997. The workshop is also now available online in several languages (see page 265).

ExxonMobil addresses eight SDGs within its sustainability activities, of which five are also in the area of community investments. These are SDG 1 "End poverty in all its forms everywhere", SDG 3 "Ensure healthy lives and promote wellbeing for all at all ages", SDG 4 "Ensure inclusive and quality education for all and promote lifelong learning opportunities for all", SDG 5 "Achieve gender equality and empower all women and girls" and SDG 8 "Promote sustained, inclusive and sustainable economic growth, full and productive employment and decent work for all".[143]

PROVIDE MATCHING FUNDS

Employees are increasingly involved in the social engagement of their companies. We saw examples such as DHL, KLM and PIMCO, where employees can spend time serving communities often during work hours, for example, a few hours to a few days per year.

Several analyses show that employees increasingly wish to work in companies that contribute to societies. In the US, according to statistics for 2017, nearly 60% of companies offered paid time off for employees to volunteer, and an additional 21% planned to offer release time in the next two years. 86% believe that employees expect them to provide opportunities to engage in the community and 87% believe their employees expect them to support causes and issues that matter to those employees.[119]
In other countries, the numbers might look a bit different as culture and approaches are different.

Another element is increasingly connected to this topic: companies financially supporting the giving of employees, as in the Reemtsma example. The same statistic revealed for the US that an estimated 2-3 billion US dollars is donated through matching gift programs annually, which has upside potential as an estimated 6 to 10 billion US dollars in matching gift funds go unclaimed per year.[119]

How does matching work? Let´s see three examples.

Bank of America has a commitment to building a culture of giving and volunteering at the company. The bank's Matching Gifts Program encourages employees to contribute to causes they care about most by doubling the impact of their charitable donations to eligible nonprofits with matching gifts up to 5,000 US dollars per employee, per calendar year. The Bank of America Charitable Foundation provides more than 25 million US dollars in matching gifts annually to employee donations.[144]

Leading metals and mining company BHP encourages employees to be active citizens in their communities and support local charitable organizations. The global Matched Giving Program recognizes the contribution of employees by matching their personal financial donations. In 2019, 4,175 BHP employees participated and contributed to 350 organizations, which received a total of 3.21 million US dollars.[125]

Since 2010, Disney has matched 71 million US dollars in employees' personal donations of time and money through the Disney Employee Matching Gifts program. The company matches financial donations of employees to charitable organizations around the world. It also provides an opportunity for employees to turn their hours of volunteer service into a financial contribution through VoluntEARS Grants.[122]

PARTNER WITH OTHERS

In many areas of sustainability, it is necessary and useful to partner in order to solve the immense challenges lying ahead. This also transfers to the sphere of social engagement where companies provide their resources to solve social problems. Partnerships are common in social engagement, but not mandatory. According to a survey related to US donation statistics,

90% of interviewed corporations indicated that partnering with reputable nonprofit organizations enhances their brand and 89% believe partnering leverages their ability to improve the community.[119]

Partnerships in strategic programs are often longer term. They can include just one partner or several, either other companies or multi-stakeholder partners. Numerous nonprofit organizations base their approaches on partnerships with corporations. SOS Children's Villages, for example, the world's largest organization focused on supporting children without parental care and families at risk, cooperates with many companies. Among them is CEWE, a leading European photo service provider. CEWE partners with SOS Children's Village at a corporate level, including several sites that cooperate with local branches of this organization.

Additionally, CEWE partners for its diverse community projects with numerous organizations and initiatives at a local level for different community projects that foster biodiversity, promote photo culture or support children and other social causes in the communities across its 14 European sites.[121]

The environmental organization WWF states that *"Partnerships play a key role in WWF's efforts to influence the course of conservation. Lasting conservation is achieved through collaboration with a range of extraordinary partners, including governments, local communities, businesses and individual donors. We leverage the strengths of these collaborations to achieve great success."*[145] The WWF offers numerous ways of partnering, including giving and cause marketing, also called cause-related marketing, which is when companies promise customers that they will give a percentage of the sales of a product to a certain cause. Often, this is a promotional activity for a certain product and time period.

For Olivela, a retailer of luxury designer fashion and beauty brands, giving is integrated into its regular business. Olivela donates a relatively high percentage of 20% of the proceeds from every purchase to different causes. These include girls' education, women's empowerment, climate action, health services and biodiversity, among others. One partner is the Global Wildlife Conservation (GWC) whose mission is to conserve the diversity of life on Earth. As part of the GWC partnership, Olivela chose to focus on *"often overlooked but highly threatened species, which are just as integral to global biodiversity as pandas, tigers and polar bears. The 'underfrogs' if you will."*[146]

Partnerships can also unite different companies for a common cause, which is very effective if pursued for a defined target group and likely in a regional area, such as in the following example. "Suited for success" is a UK-based program founded on collaboration. Its initial idea is to provide appropriate interview clothes for homeless and long-term unemployed people trying to return to work. Suits, shoes, ties and other items necessary for an interview are collected and distributed. A job interview coaching service is also provided. The law firm Gowling WLG provided pro bono intellectual property, real estate and planning advice early on to help Suited for Success become a registered charity. Now, the organization ensures that more companies with their particular skills and resources contribute to the cause. First Impressions, an international image consultancy, offers image skills and professional development training. Francesco Group Birmingham Academy offers free haircuts and styling appointments. Uber helps unemployed men and women get to their job interviews without financial stress or the need to navigate their way to a new place by offering them free Uber rides. ⤴ www.suitedforsuccess.co.uk/about-us

HOW TO DO GOOD, WELL

We have seen that many companies "do good" far beyond their business. They provide significant resources to social and environmental causes across multiple communities globally. Yet, public appreciation for this engagement is often restrained. Frequently, companies face significant mistrust or opposition, even when it comes to their social engagement. Corporate executives often don't really understand why they are confronted with criticism when they publicize their good intentions and significant investments. Why does this happen, and how can it be addressed?

SOME CRITICAL QUESTIONS ON ENGAGEMENT

To critical readers, the last previous pages may have sounded very positive and a little one-sided, and critical questions could be raised about some of the approaches of social engagement described. The typical reservations of greenwashing (see chapter 5) also apply to social engagement—maybe even more so as it is about doing good. Moral standards appear to be even higher in such an arena.

To give an insight into the thoughts behind typical reservations, let's raise some critical and provocative questions about social engagement. Not all questions reflect my professional or personal point of view, but they are often alluded to.

– If Kellogg's donates its cereals to be distributed among poor people: Is the overall intention to get new customers hooked? Is it truly a good meal for hungry people, considering nutritional value and sugar content?

– If Apple engages by giving its products to schools and teachers in poor communities—don't they just raise the desire of low-income people to get Apple's newest products? And is it credible on the one hand to spend millions of dollars on school programs by donating products while across the supply chain precarious conditions prevail in mining or when workers treat e-waste? (see pages 284ff.)

– When Google is frequently mentioned as a company trying to avoid paying any tax whatsoever, how credible is such immense social engagement?

– If luxury retailer Olivela donates 20% of its sales to different causes, don't they just stimulate more sales of items people don't really need? Why do they leave out critical issues such as unfair wages for clothing? Could the money not be given directly to people in need? Or could prices be reduced? And isn't the whole luxury industry promoting and cementing unhealthy consumption patterns, disparity and inequality across the world?

– If motorcyclists raise money for kids with brain tumors through the Honda Riders Club—is it justified to worsen the climate problem with the extra emissions that result from riding? Is Honda not greenwashing—that is, distracting from its original challenge of climate impact through helping sick children?

– If FrieslandCampina supports the dairy industry in other countries, can this be labeled social engagement? Is it not just another way of investing in future markets? Don't they misuse a social cause to get access to new customers and "buy" their reputation from local influencers?

– Are the millions of dollars companies spend on social engagement not a disadvantage for shareholders? Shouldn't the shareholders themselves decide which causes they want to support at a personal level?

– Are motivations such as the one described by Honda, IKEA and L'Oréal truly credible? How does it align with corporate misbehaviors that often occur across all kind of companies?

– If Goldman Sachs employees spend millions of dollars on social causes but, at the same time, in their daily job nurture a fragile financial system that affects many people around the world negatively, is it truly acceptable? Is it like selling indulgence?

– Is it good if ExxonMobil employs local staff and invests in charity but at the same time contributes significantly to climate change and loss of biodiversity?

– How does doing good go hand-in-hand with companies laying off people, even when they are doing economically well?

– If meat producer JBS helps people in the pandemic in Brazil, but the business model of its daily operations contributes significantly to destroying rainforests in Brazil and to climate change worldwide, how sincere and relevant is the disaster relief?

Some questions and underlying concerns may resonate more than others, as perceptions and expectations are highly individual. They depend on experiences, impressions and personal preferences, among others.

Basically, two major concerns underlying most of the questions raised above can be detected:

1) Whether the company is trustworthy and, specifically, its motivations

2) If social engagement is used to compensate for shortcomings in other, more relevant, areas

Let´s look at both. This will not provide answers to each of the questions but it will help companies reflect more on the credibility of social engagement. Also, I believe, it also will gradually help to better serve the causes.

NAVIGATING STRATEGIC ENGAGEMENT

There is a fine line that businesses have to meet. When doing good, good intentions are mandatory. A hidden agenda is rarely accepted. Moral standards towards companies, as well as expectations about their contributions and their integrity, are high. The general mistrust many companies face regarding an attitude to "just sell stuff" transfers to social engagement. Being perceived as citizens does not happen fast. Trust grows only slowly over time. It is necessary to be aware of this, consider it in program design and frequently question if credibility exists—which includes inviting others to give their feedback.

From my experience, brand-driven companies, in particular, have a harder time navigating this area successfully, as described earlier. They transfer the strong brand image and business mechanisms 1:1 to sustainability and social causes, often supported by their marketing agencies. But many social causes are perceived as too important and serious to be addressed by the brevity of a regular branding campaign. It is good to question motivations and intentions.

As it is a comparably new arena for companies—as it still is for many non-profit organizations—intensive engagement, in contrast to short-term activity, can be crucial. It gives actors a chance to meet as people, connect to a common cause, do good together. True commitment, openness about motives, long-term orientation, humbleness and an appropriate communication—rather more reserved than very bold—are critical success factors.

Several years ago, I had the chance to accompany a big citizenship project for a few years. It was an unusual cooperation between a major energy company and an environmental organization with the goal to engage in pre-school education on energy topics and sustainability. It took over a year to build trust between both partners and to agree on the terms of the project. As it targeted young children, the approach was even more delicate to navigate. Several initial marketing intentions of the company—apparently the bigger partner as it was also the money provider—had to be adjusted over time to make it a credible, impactful project. It turned out to be a major organizational change program and learning journey for both partners. Compromising over and over again in favor of the common cause and truly valuing the people on the other side, in addition to being open to learn from each other, eventually made the project possible. These side-effects are as important as the impact generated.

In summary, strategic social engagement is not something where you should expect quick campaign success. It is about honest interest and true commitment to the cause. Citizenship should be carefully planned and executed as several potential pitfalls are on the way.

BUILDING ON A SOLID FOUNDATION

Social engagement is great, but it is an extra for a company. I want to be very clear about this: By no means, can engagement cover or compensate for the homework a company has to do regarding sustainability in its core business and value chain. It does conflict with basic ethical topics, such as tax evasion and corporate scandals, and even when companies have to make hard decisions such as job cuts. Social engagement is a valuable yet delicate contribution. Compensation is usual not the intention, but a broader holistic understanding and appropriate actions are often lacking.

This is a long-term process. A company doesn't have to be 100% "ready" from a sustainability or moral standpoint, which would be unrealistic anyhow. But any major strategic social engagement should be built upon a solid, credible foundation of material sustainability activities. This is our fifth criterion (see page 237). Get your house into a basic order before engaging in social causes at a strategic level. Or, at the latest, start and progress while you are engaged. This holds especially true for companies from critically observed industries and, as stated, for brand-driven companies, especially multinationals. It is also relevant for any company that engages with corporate resources in topics close to the core business.

That became clear in the project mentioned above: An energy company— seen critically by the public— engages in energy education for children— close to the core business and delicate. The sustainability level of the company at the start of the project was not very advanced. It improved over the course of the project but ultimately did not sustain the project beyond its pilot period of four years of operations. There has to be step-by-step advancement in both areas—core sustainability and additional social engagement. Certainly, other reasons also contributed to not extending the project.

From my experience, in practice, sustainability advancement and strategic citizenship often do not go hand-in-hand. They are often treated as separate activities, also organizationally. For most big and multinational companies, the alignment has to improve in the long term. This includes solid sustainability activities along the major material issues, an aligned strategic social engagement approach, as well as aligned structures and processes. Let's finish with a short checklist to assess this.

It is a heartfelt issue to see more impactful social engagement from companies due to the great potential lying in this field. I also advocate for more public appreciation of company efforts and, at the same time, a constructive discourse on critical questions about engagement in balance with sustainability performance.

 **TOOLBOX**

CHECKLIST: IS CORPORATE CITIZENSHIP ALIGNED WITH SUSTAINABILITY?

1. Are you in an industry generally seen critically or in an industry under public scrutiny? (see page 151)

2. Do you lack a solid sustainability strategy along material topics with specific targets?

3. Does your corporate social engagement lack the criteria for strategic engagement? (see page 237)

4. Were you wondering about some of the questions raised above, about the credibility of social engagement and why anybody should care about such topics?

5. Have you received criticism in the past for your sustainability activities, especially for not addressing the critical topics of your industry very well?

6. Are you communicating your social engagement extensively and proudly?

7. Are personnel structures and reporting lines for social engagement and sustainability activities separated?

If you answered "Yes" to more than 3 of the above questions, it is recommended that you get a status check from an expert to receive input for mid-term alignment.

APPLICATION

1. ——— Test your knowledge!
   True or false?
   a) Corporate citizenship, social engagement, and
   community contribution are often used synonymously.
   b) Matching funds double the money employees
   spend on social causes.
   c) Companies just do good to sell more products.
   d) Linking social engagement to the SDGs is voluntary.
   e) Corporate sustainability does not intersect with
   social engagement.

2. ——— Which examples in this chapter appeal to you? Why?

3. ——— Which examples in this chapter raised questions regarding credibility? Why? How could it
   become a credible contribution?

4. ——— Think of your personal engagement for society such as donating money, devoting time or other resources to social causes, or advancing your own project.
– What are your motivations?
– How do you approach it?
– In which way does or should corporate engagement towards social causes differ from personal engagement as you know it?

5. ——— Discover with kids!
In BASF's Virtual Lab, young scientists ages 8 to 12 can choose from a vast array of interactive chemical experiments and learn things like how to clean dirty water, why laundry gets discolored, and how solar energy and photosynthesis work. The Virtual Lab is available in several languages.
Landing page (from there, select languages):
basf.kids-interactive.de

REVISING GLOBAL DYNAMICS

*I love to travel. If there is one thing that really
ruins my carbon footprint, it is traveling.
I am not saying this light-heartedly. And, although
I offset the emissions of my flights, plan consciously,
and avoid flights whenever possible, I still leave a footprint
that would not be there if I didn´t travel. At the same time,
traveling has shaped my personality, my work, and my view
of our beautiful world and its people extensively.*

*Travel taught me vividly at a professional, an emotional
and personal level: that how business dynamics are set up
can leave either a good or bad footprint on people and
nature around the world. And what companies do is often
closely connected to what consumers want or don´t want.
Can new paths be found?*

Our world is interdependent as value chains are highly globalized. Climate change and loss of biodiversity, for example, occur on a global scale but have specific local impacts and hit some individuals more than others.

UNEQUAL DISTRIBUTION OF BURDENS

Benefits and burdens are not equally distributed across the world. Per capita consumption growth is spread unevenly across the world with rich countries consuming above average. Developments like climate change, pressure on natural resources and demographic change affect some regions, nature and people disproportionally. Here are a few selected examples as an illustration.

> **TIP**
>
> How, where, and by whom are our products made?
> An insightful BBC series on the history and dynamics of the global economic cycles for eight products, including flowers, bicycles, semiconductors, paper and whiskey.
> ⌁ tinyurl.com/y5trjmzx

– Climate change affects countries and people very unevenly. The UN-based FAO analyzed the countries most vulnerable and at risk of climate change. Coastal regions of sub-Saharan Africa and small island states in the Pacific such as Benin, Kiribati, Liberia, Mauritania, Mozambique, Sierra Leone, Solomon Islands and Togo all have the highest risk, together with Cambodia and Haiti. They are socially and economically vulnerable, but what affects them the most are the significant negative effects of climate change.[147] This is highly contrary to the main originators of climate change (see page 49).

– The 2016 Global Climate Risk Index showed that between 1995 and 2014, more than 525,000 people died as a direct result of approximately 15,000 extreme weather events. The total amount of losses incurred was over 2.97 trillion US dollars. The World Bank also estimates that by 2030, there could be 325 million people trapped in poverty and vulnerable to extreme weather events in sub-Saharan Africa and South and Southeast Asia.[148]

– The loss of coastal habitats and coral reefs, which reduces coastal protection, increases the risk from floods and hurricanes. This affects the lives and property of the 100 million to 300 million people living near the coasts.[29]

– Resource exploitation often takes place in fragile areas. The consequences of agricultural expansion were felt primarily in some tropic regions, home of the highest diversity on the planet. The loss of millions of hectares of tropical rain forests for cattle ranching occurred in Latin America and for palm oil plantations in South-East Asia. (see page 95)

– More than 2 billion people rely on wood fuel to meet their primary energy needs, and an estimated 4 billion people primarily rely on natural medicines for their health care.[29] These people are especially affected by higher burdens on wood and the loss of biodiversity.

– Commodity dependency is heavily concentrated within developing countries, with 85% of the least developed countries, 81% of landlocked developing countries and 57% of small island developing states being commodity-dependent, compared to just 13% of developed countries, according to a report of the UN trade organization UNCTAD. Multi-nationals that heavily source commodities such as cobalt, cotton, palm oil, coffee and oil, among others, can be shown to have a significant impact in developing countries through their supply chains.[128]

– Food insecurity, missing access to clean, safe drinking water or sanitary facilities, child mortality and other challenges remain issues for regions, especially in Africa, Asia and Latin America. Environmentally-based health burdens, such as air and water pollution, are more prevalent in least developed countries.[29]

– Least developed countries, often rich in and more dependent upon natural resources, have suffered the greatest land degradation, have also experienced more conflict and lower economic growth, and have contributed to environmental outmigration by several million people.[29]

– Less than half of the global population is covered by essential health services.[105]

– Fewer than one in five people use the internet in the least developed countries.[105]

It shows an overall global imbalance that affects several severe social and environmental challenges, often connected to business activities.

THE VULNERABILITY OF TOURISM

Let´s take this to a practical level with the example of an industry that is highly impacted by global dynamics. The travel and tourism industry plays an important role in the global economy. In 2019, international tourist arrivals reached 1.46 billion, with total tourism receipts reaching 1.478 trillion US dollars[149], or equal to about 4 billion US dollars per day. In 2018, this industry generated 10.4% of world GDP and a similar share of employment. In the decade ahead, tourism's contribution to GDP is expected to rise by nearly 50%.[150]

In May 2020, tourism dropped by 98% and was expected to decrease by a total of at least 56% for the year 2020 as a whole, according to an estimate of the UNWTO early in 2020.[151] However, as hard as Covid-19 has hit the industry, its vulnerability is not new or a result of external shocks. Instead, it illustrates the interdependence of economic, environmental and social factors through the way business is done.

Take the example of the Maldives, an epitome of paradise, consisting of over 1,000 islands, most of them uninhabited. Tourism accounts for 87% of the country´s exports and has grown by an annual rate of 10% over the past ten years.[152] Economically, tourism is highly important for the country.

In 2009, the president of the Maldives at the time held an "underwater cabinet meeting" to draw international attention to the fact that his country was severely threatened by climate change. Rising sea levels—an already observable consequence of climate change—was on its way to flooding most of its islands and endangering people´s lives and ecosystems. A decade later, not much has changed. Today, officials are reminding

LOSS OF BIODIVERSITY

It is a topic well discussed among experts but publicly still widely under the radar: Loss of biodiversity refers to either the ongoing extinction of species at a global level or the local reduction or loss of species in a given habitat.

Most species currently extinction-threatened are so because of human activity. In the last 40 years, we have lost 52% of the planetary bio-diversity and 58% of vertebrates on land, sea and air, according to the Global Assessment Report on Biodiversity and Ecosystem Services released in 2019. Below is a summary of a few of the underlying dynamics identified by the report.[29]

– Land use change is the direct cause and has the greatest impact on terrestrial and freshwater ecosystems. It is driven primarily by agriculture, deforestation and urbanization, all of which are associated with air, water and soil pollution.

– 25% of the globe's greenhouse gas emissions come from land clearing, crop production and fertilization, with animal-based food contributing 75%.

– Direct exploitation of fish as well as seafood has the largest impact on the oceans. Marine plastic pollution has increased tenfold since 1980, affecting at least 267 species, including 86% of marine turtles, 44% of seabirds, and 43% of marine mammals. This can affect humans through food chains.

– In terms of direct exploitation, approximately 60 billion tons of re-newable and nonrenewable resources are being extracted each year. That total has nearly doubled since 1980, while the population also has grown considerably. The average per capita consumption of materials, such as plants, animals, fossil fuels and ores, as well as construction materials, has risen by 15% since 1980.

– Cumulative records of alien species have increased by 40% since 1980, associated with increased trade and human population dynamics and trends.

– Climate change, pollution and invasive alien species have had a lower relative impact on loss of biodiversity to date but are accelerating.

Regarding policy progress in countries, the United Nations stated in 2020, that only a third of 113 countries were on track to achieve their national target to integrate biodiversity into national planning.[105]

An article that concisely explains and illustrates the major environmental impacts of biodiversity loss, is found on the encyclopedia Britannica: www.britannica.com/science/biodiversity-loss

us again of the vulnerability of their country, its people and nature and a responsibility to address such topics in financial partnership. Watch the short video: tinyurl.com/y6xmt4sv

Another global topic that has turned into an emergency in this country is waste. The Maldives are now the area with the world´s highest level of microplastic on beaches and in waters near shore. Not only that, but they are also one of the countries with the worst per capita mismanagement of waste, according to a National Geographic Article.[153] Apparently, plastic reaches the islands as a result of global plastic waste in the oceans.

Another issue is the fact that a single tourist produces almost twice as much trash per day as a resident of the capital city of Malé, and five times as much as residents of the other 200 populated islands, according to government statistics released in the context of a World Bank project on waste management.[154] Maldives is also a developing nation lacking local

TIP

The website of the Travel and Tourism Competitiveness Report of the World Economic Forum provides rankings, detailed analysis of regions and country profiles. Some sustainability related indicators are embedded into a broad set of competitive factors.

⤳ tinyurl.com/y6mmterx

manufacturing. Most consumer goods that provide for tourism must be shipped or flown in. Due to its natural constitution and the current environmental and social effects of business activities, the Maldives is a case of an especially vulnerable nation.

But the industry has general challenges to solve, one of them being a high carbon footprint. Tourism-related transport emissions represented 22% of all transport emissions in 2016. At that time, they account for 5% of all man-made emissions and are projected to increase to 5.3% by 2030, according to the UNWTO.[151] Another study showed that, in 2013, there was already a share of 8% of total global greenhouse gas emissions, which included not only transport but also tourism-related food consumption.[29] A rising economic importance and the existing conditions—many poor tourist destinations depend heavily on tourism while being among the world´s poorer countries—require future scenarios for more sustainable industry development.

What would a more sustainable path in travel and tourism look like? I chose as a vision for the trajectory a quote from TTS, a global provider of innovative solutions for the travel and tourism industry. It asks for a new business model for the travel and tourism industry: *"The travel industry should be part of helping build up the local communities, be part of the culture, and support the people that the industry comes in contact with, instead of merely descending on them."*[155]

TREATING CLOTHES AS COMMODITIES

The apparel industry impacts millions of people in many, often poorer, countries around the world. Over the past decade, several developments have deepened the distortions in the industry. Decisions by actors across the textile value chain were mainly taken from an economic standpoint. Consumers latched onto and nurtured the unhealthy dynamics. The challenges in the industry that had existed for many years became even more apparent. It might be time to rethink how we want to shape the future of this industry.

UNHEALTHY BUSINESS DYNAMICS

A McKinsey study summarizes the trends taking place the past few years in the apparel industry: *"Thanks to falling costs, streamlined operations, and rising consumer spending, clothing production doubled from 2000 to 2014, and the number of garments purchased each year by the average consumer increased by 60% ... businesses have enabled shoppers not only to expand their wardrobes but also to refresh them quickly. Across nearly every apparel category, consumers keep clothing items about half as long as they did 15 years ago. Some estimates suggest that consumers treat the lowest-priced garments as nearly disposable, discarding them after just seven or eight wears. Shorter lead times for production have also allowed clothing makers to introduce new lines more frequently. Zara offers 24 new clothing collections each year; H&M offers 12 to 16 and refreshes them weekly. Among all European apparel companies, the average number of clothing collections has more than doubled, from two a year in 2000 to about five a year in 2011."*

More clothes, lower prices, less durability. The trend is alarming, as developing countries catch up, as the article continues: *"Shoppers have responded to lower prices and greater variety by buying more items of clothing. The number of garments produced annually has doubled since 2000 and exceeded 100 billion for the first time in 2014: nearly 14 items of clothing for every person on Earth. While sales growth has been robust around the world, emerging economies have seen especially large rises in clothing sales, as more people in them have joined the middle class. In five large developing countries—Brazil, China, India, Mexico, and Russia—apparel sales grew eight times faster than in Canada, Germany, the United Kingdom, and the United States."*[156]

During Covid-19, this consumption cycle was interrupted, as lockdowns across the world affected shops and malls as well as overall spending. A direct consequence was millions of garment workers around the world missing out on their regular wages, or not being paid at all, according to a report by the advocacy organization Clean Clothes Campaign. The report *"Un(der)paid in the pandemic"* analyses the nonpayment of wages to garment workers during the months of March, April and May 2020 resulting from order cancellations by apparel brands, unpaid leave, and state-sanctioned wage cuts during the Covid-19 crisis. Across South and South-east Asia, garment workers have received 38% less than their regular income. In some regions in India, this number rises above 50%. Extrapolating these findings to the global garment industry, a conservative guess of wages lost by garment workers worldwide, excluding China, for the months of March, April, and May 2020 would amount to between 3.19 and 5.79 billion US dollars. ◁ tinyurl.com/yyxvs6mw

In Cambodia, production was temporarily stalled for 450 fabrics from the apparel industry, leaving 150,000 workers without a job. The Asian Development Bank is estimating that about 390,000 jobs will be lost in the country in 2020 due to the pandemic.[157]

But is it just a pandemic-related problem? Many industries and millions upon millions of people across the world have been affected during the crisis. The systemic anomalies hit people hard that even in normal times were already vulnerable. The global apparel industry is based on fast consumption in rich countries and production at unsustainable levels in poorer countries. Some see it as fashion at a commodity-level, others as crucial for survival, as we will see.

THE REAL WAGE ISSUES

The Fair Wear Foundation, which strives to improve labor conditions in garment factories, monitors different manufacturing countries by looking at audits and complaints. It reported violations in multiple countries in 2019 that were often related to wages or other issues for workers, most of them women.

TIP

Two videos explain the global textile supply chain and some of its major issues.
tinyurl.com/y6ngrfzf
tinyurl.com/y6o9u7gs

For example, in Bangladesh, the most common labor violations relate to low wages, forced excessive overtime, no freedom of association, poor social dialogue, and harassment. Complaints were often related to dismissals and not receiving the correct compensation. Insufficient wage payments, excessive overtime and unpaid overtime were reported from China. In Myanmar, the most common labor violations are related to low wages, excessive and forced overtime, child labor, no freedom of association, and the lack of a constructive and healthy dialogue between workers and management. Although the minimum wage was increased, many workers complain that their cost of living has also risen substantially, their overtime is often cut, and they miss out on some bonuses due to unrealistic production targets. In Vietnam, labor rights issues are related to wages, as well as to excessive overtime. This includes exceeding the legal limit of overtime hours, working during breaks, continuing work after working hours without payment and recording work hours, and working on Sundays. In some cases, workers found it impossible to refuse overtime.[158]

A different issue has been reported for many years from Uzbekistan and Turkmenistan: In these major cotton-growing countries, forced labor supported by the government continues to be a problem. Non-governmental organizations and multi-stakeholder initiatives have been fighting against this for many years as shown on this website: www.cottoncampaign.org

Companies and brands in the textile industry usually use the local minimum wage in the respective production country as a guide. This is a typical practice in businesses and countries across the world. For a number of countries producing for the fashion industry, the Clean Clothes Campaign publishes the gap between the national minimum wage and the living wage that would allow a decent living. In countries such as Bangladesh, Ukraine, Serbia and Bulgaria, the minimum wage is less than a quarter of the living wage. In Cambodia, India, Indonesia and Turkey, it is a bit more. But in no country does the minimum wage reach even half of what would be considered a living wage.[159]

The human right to a living wage is defined in the UN Universal Declaration of Human Rights. The International Labour Organization states: *"One of the fundamental human rights is the right to a just remuneration that ensures an existence worthy of human dignity. The preamble to the Constitution of the International Labour Organization identifies the provision of an adequate living wage as one of the conditions for universal and lasting peace based on social justice ... Although there is no universally accepted amount that defines such remuneration, it can be described as a wage from full-time work that allows people to lead a decent life considered acceptable by society."*[160]

TIP

A website monitoring companies and brands for their performance regarding wages and other aspects.

www.fashionchecker.org
www.fairwear.org/brands

Transparency with respect to how companies address labor conditions and wages is increasing. The wage issue applies to most fashion brands—just as much to luxury labels as to discount fashion. Although the payment of a living wage is not solely influenced by the brands (but also local governments, and others), the pressure for better wages for garment workers will increase.

This is not only a topic in Asia, but also increasingly in Eastern Europe. Romania, North Macedonia and Bulgaria have all been audited by Fair Wear Foundation and the results show that low wages, unpaid overtime, unpaid holidays and fire safety concerns continue to be the most prominent rights violations. Factories are still far from paying a living wage and sometimes do not even pay the minimum wage.[158]

The Clean Clothes Campaign frequently publishes country profiles on its website showing the country's labor conditions. European countries, such as Moldavia, Hungary, Croatia, Romania and the Czech Republic, are also producing clothes and shoes for multiple brands—many of which are highly regarded.

Romania is the country in Europe with the most garment production. It is estimated that 400,000 people work in the garment and footwear industry— 300,000 working formally and the rest informally. According to the country report, garment workers earn a (monthly) minimum wage of about 230 euros after tax. Workers reported that they would need triple their current wages for a decent life to be possible.[161]

Romanian garment factories are producing for brands such as Naf Naf, Max Mara, Sisley and Benetton. At the time of researching this topic in the summer of 2020, Naf Naf was having a large sale with 60% off. Including this discount, a shirt would cost between 18 to 26 euros. According to Fair Wear, from the purchase of a 29 euro shirt, an average of 0.18 eurocents goes to the tailor in the garment factory.

The business model for targeting maximum consumption based on wages that don't allow a decent living for millions of garment workers will be a challenge to address collectively. It requires enormous efforts on the political levels of the respective countries (rising minimum wages, controlling local factories), on producing brands (enforcing decent working conditions, paying higher wages and increasing order reliability) and consumers (overall less and more conscious consumption).

From the environmental perspective, the question remains, too: How can dynamics be revised?

The water footprint of a single T-shirt is high, requiring 2,700 liters of water to produce on

TIP

The Garment Worker Diaries provide first hand insights, a podcast and a call to action.

⊲ workerdiaries.org

average. This equals enough drinking water for one person for 2.5 years.[162] Shouldn't a single T-shirt be much more valuable than the disposable, easy-to-buy, throw-away commodity it has often become nowadays?

Other environmental impacts of the apparel industry are addressed in several other passages. Organic cotton as a percentage of all cotton produced worldwide is low at 0.7% (see page 109). Only 1% of all textiles are recycled (see pag 339). The issue of microplastic release (microfibre pollution) gains increasing attention. All of these ecological impacts resulting from the apparel industry are only slowly building public awareness.

9.02 ——— SMARTPHONES BEHIND THE SCENE

In comparison to the apparel industry, the smartphone industry is very young. The industry only just celebrated its 10th anniversary in 2017, and many of us can still remember a time without mobile phones. Here was a terrific chance to build up an industry from scratch in the 2000s. How did we do?

On the 10th anniversary of the smartphone, Greenpeace released a report on just that question. A short video, accompanying its release, takes stock and acts as a humorous yet sincere wakeup call—not only for the industry but also its customers. ✐ tinyurl.com/y46k7xul

We will look at the lifecycle of smartphones from sourcing, to manufacturing, to recycling.

DO YOU REALLY NEED A SMARTPHONE?

First of all, it is important to understand that an entire industry was built on the narrative of continuous consumption. Market numbers are impressive. In 2020, 3.5 billion people in the world used a smartphone; according to Statista, the number has grown from 2.5 billion in 2016. China, India, and the United States have the highest number of smartphone users.[163]

TIP

Diverse statistics around smartphone use and trends— such as 90% of smartphone time is spent on apps and the average time spent on mobile devices is 2 hours and 51 minutes per day.

tinyurl.com/y2ro4flb

In 2019, 1.372 billion smartphones were produced. In 2020, due to Covid-19, the worldwide smartphone market is expected to decline by 11.9%—the largest year-over-year decline in its short history. Nevertheless, it will still reach 1.208 billion units, according to the International Data Corporation (IDC).[164] Tablets accounted for 144 million units worldwide in 2019, according to IDC.[165] During the Covid-19 crisis, IDC reports that consumers spent less on smartphones and significantly more on other technologies, such as PCs, monitors, and tablets, to meet the demands of mandatory work from home and distance learning.

The tremendous growth of the smartphone market was not only a result of first time, new users. On the contrary. People in rich countries very frequently "upgrade" to the newest model. In fact, a new smartphone is purchased every 2.2 years on average. "Get the latest" is in vogue and encouraged by the marketing of basically all mobile device manufacturers and telecommunication providers. Offers such as get a phone for 1 euro, connected to a contract, are an invitation for frequent upgrading. The habitual invention of new devices (tablets, smartwatches etc.) for all occasions nurtures the business cycle for new devices. The ecological and social footprint are only slowly coming to the attention of consumers and the public.

"Do you really need a smartphone or do you just want one" was, for example, the provocative title of a factful article that describes, among other things, why our smartphones are *"a raw material treasure chest"*. They contain numerous metals, chemicals and alloys, which are mined across the world. The conclusion, based on the effects, was that we should resist the "need" to constantly upgrade our smartphones, and that the industry has to transition towards circular solutions and repairs instead of creating demand for new devices.[166] The same conclusion was drawn by the

aforementioned Greenpeace report released in 2017. Its title: *"From smart to senseless—the global impact of 10 years of smartphone"*.[33]

MINING FOR SMARTPHONES

The Greenpeace report shows that about 60 different materials are used in a typical smartphone—as well as how the amount of raw materials extracted for smartphones has increased over the past ten years. For example, an average smartphone contains 22.18 grams of aluminum. In the period from 2007 to 2017, this amount summed up to 157,478 tons for all smartphones produced. Copper used in wiring in an average smartphone accounts for 15.12 grams and totaled 107,352 tons in that same period. Cobalt, as part of battery production, usually amounts to 5.38 grams. It has accounted for 38,198 tons since 2017.[33] Among the raw materials extracted for smartphones are rare earths.

> **TIP**
>
> Despite of their name, rare earth elements are relatively plentiful but usually scattered and found in low concentrations, due to their geochemical properties. This means many raw materials have to be mined to be extracted. A video shows precious pictures of 16 of the 17 existing rare earth metals. The photographer shows the natural fragility of the metals.
> vimeo.com/431119901

According to the IPBES 2019 study on biodiversity, the overall impact of mining (not only for smartphones) has significantly increased. And though its economic contribution to the mining countries is significant, the effects, especially on the environment, are severe: *"All mining on land has increased dramatically and, while still using less than 1% of the Earth's land, has had significant negative impacts on biodiversity, emissions of highly toxic pollutants, water quality and water distribution, and human health. Mined products contribute more than 60% of the GDP of 81 countries. There are approximately 17,000 largescale mining sites in 171 countries, with the legal sites mostly managed by international corporations, but there is also extensive illegal and small-scale mining that is harder to trace, and both types of sites are often in locations relevant for biodiversity."[29]*

About 83% of the global mining workforce, or 40.5 million people, as well as the surrounding communities, rely on artisanal and small-scale mines, according to Fairphone.[167] Working conditions are often precarious. Global mining is under the radar of public awareness, and so are some issues potentially associated with it. Child labor linked to smartphone battery making has become a public concern, specifically within the cobalt supply chain in the Democratic Republic of Congo, according to a 2016 Amnesty International (AI) Report. Years ago, international organizations had already pointed out the problem of children mining cobalt in the DRC. AI cites a report from UNICEF estimating that, in 2014, approximately 40,000 boys and girls worked in mines across southern DRC, many of them involved in cobalt mining. This type of work involves long hours, frequent lifting and transporting of heavy loads, and working without protective equipment such as gloves and face masks. Such conditions frequently expose people to high levels of cobalt, which can result in a potentially fatal lung disease.[168] Several recognized multinational corporations were accused of claiming zero tolerance for child labor in their supply chains, but at the same time failing to effectively trace and enforce it—often because issues occurred in the second or third layer of the supply chain, or because political situations were difficult.

> **TIP**
>
> A factful article and, for more in-depth information, the full Amnesty International report "This is what we die for: Human rights abuses in the Democratic Republic of the Congo power the global trade in cobalt".
>
> ⚡ tinyurl.com/y68yfocm
> ⚡ tinyurl.com/y4mzvfff

Additionally, the topic of conflict minerals arose. These currently include the metals tantalum, tin, tungsten and gold, which are derivatives of minerals often referred to as 3TG.[169] The issue of conflict minerals emerged with the Dodd Frank Act, a 2010 amendment to the US Exchange Act. It was enacted to address armed groups' exploitation of and trade in 3TG minerals, which is partially financing the conflict in the Democratic Republic of Congo. Now, 3TG minerals, regardless of where they are extracted or whether they directly or indirectly benefit armed groups in the included countries, are treated by this regulation as conflict minerals. The OECD Due Diligence Guidance for Responsible Supply Chains of Minerals from Conflict-Affected and High-Risk Areas has an even broader scope and covers all minerals, not only 3TG.

A BIG FOOTPRINT IN MANUFACTURING

The challenges in the sourcing phase of the smartphone are still relatively little known. But the manufacturing phase is significant from an environmental standpoint, too, as the Greenpeace study details: *"Electronics manufacturing is highly energy-intensive, and its energy footprint is growing significantly, as the volume and complexity of our electronics devices continue to expand. Various lifecycle analyses find the manufacturing of devices is by far the most carbon-intensive phase of smartphones, accounting for nearly three-quarters of total CO_2 emissions. Since 2007, roughly 968 TWh has been used to manufacture smartphones. That is almost as much electricity for one year's power for India, which used 973 TWh in 2014."* [33]

> **TIP**
>
> The Responsible Minerals Initiative was founded in 2008 to pave the way towards a more sustainable supply chain. Today it comprises more than 380 companies and associations from 10 industries. Read about The Responsible Minerals Assurance Process, a third-party audit verifying smelters and refiners sourcing responsibly in line with current global standards.
>
> tinyurl.com/yxpvdleg

The Greenpeace report challenges the current business model for both manufacturers and service providers, which relies on the frequent replacement of devices. There is a demand for alternative solutions, such as the use of recycling materials, a slower replacement cycle, and manufacturing with 100% green energy.

Fairphone, a Dutch enterprise launched in 2013, takes off exactly from that point and provides a different business model to the smartphone industry. It approaches the big industry challenge: longer-lasting devices, modular and repairable design, more recycled materials. Additionally, Fairphone strives for transparent and fair supply chains. A promising story of a "Growing David" (see page 189).

AND WHEN IT'S OVER?

Between 2010 and 2019, electronic waste worldwide grew by 38%.[105] The yearly amount of e-waste is estimated to be nearly 50 million tons.[170] This is roughly equal to the weight of 25 million cars and more than 6 kilograms of e-waste per year for every person on the planet. Of the 42 million tons of e-waste produced worldwide in 2014, about 3 million tons originated from small IT products like smartphones.[33]

In the European Union, only half of the electronic waste is recycled. The other half goes to dumps and incinerators.[171] Based on figures from 2010 to 2019, the United Nations calculated that less than 20% of all electronic waste is recycled worldwide.[105] Fairphone estimates that less than 10% of smartphones are returned for recycling.[172] In 2014, German households possessed about 200 million smartphones no longer in use that are full of precious materials that could be recycled. The number doubled since 2015.[173]

> **TIP**
>
> "Changing the electronics industry from the inside" is the vision of Fairphone. The Impact Reports of Fairphone show what is possible in the industry. The social media channels show pictures and stories of the smartphone supply chain.
>
> fairphone.com/en
> instagram.com/fairphone

According to the Greenpeace report, in 2014, *"less than 16% of global e-waste was estimated to be recycled in the formal sector. Much of the rest likely went to landfill or incinerators, or was exported where dangerous informal disassembly operations threaten the health of local communities."* [33]

The United Nations "Basel Convention on the Control of Transboundary Movements of Hazardous Wastes and Their Disposal", entered into force already in 1992. The Convention prohibits hazardous waste from being exported to less developed countries. Most countries in the world have committed to the Convention; the United States has signed but not ratified it.

TIP

BAN provides in-depth reports on e-waste exports on its website.
⌁ tinyurl.com/y2af9tad

The non-governmental organization Basel Action Network (BAN) has been active since 1997 as a watchdog advocating for better enforcement. It analyzes the practices of countries (such as the US, Canada and Australia) and companies (such as computer manufacturer Dell and recycler e-Tech). Through the e-Trash Transparency Project, BAN verifies the paths of electronics via GPS-based tracking devices and disclosed several exports.

In accordance with the Basel Convention, German legislation prohibits the export of electronic waste to countries in Asia or Africa. In the past, rumors circulated that amounts of more than 150,000 tons of e-waste were being shipped abroad, some of it illegal. One reason is that, while the export of waste is prohibited, the export of second-hand articles is allowed. And although it might seem useful to give outdated smartphones, washing machines and printers a second life in other countries, these products often eventually end up in those countries' dumpsites—without proper recycling and often precarious working conditions.

Ghana was estimated years ago to be earning between 100 to 250 million US dollars per year with the recycling of e-waste and having 200,000 people working in its recycling industry.[174] In Nigeria, up to 100,000 people work in the informal e-waste recycling sector, and over half a million tons of discarded appliances are processed in the country every year.[175]

In Accra, Ghana's capital, lies Agbogbloshie. This is the largest e-waste dumpsite in Africa—and maybe even the world. Some estimates say that about 60,000 to 80,000 people work—and many live—at this recycling site. They do so among huge amounts of household and electronic waste: refrigerators, washing machines, hairdryers, automotive scrap, smartphones, computers, wires, and other items, from "somewhere" in the world.

The ban on hazardous materials in the European Union according to RoHS compliance (RoHS = Restriction of Hazardous Substances) helps to restrict hazardous materials in electrical and electronic products, among them lead, mercury, cadmium and phthalates. This reduces pollution of landfills and dangerous occupational exposure during manufacturing and recycling for these substances.[176] But still, many of the current practices at dumpsites like Agbogbloshie pose tremendous challenges.

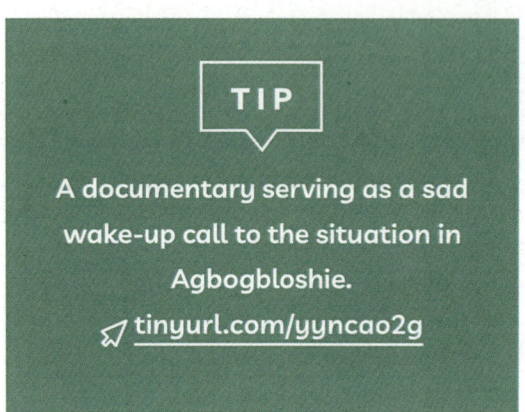

TIP

A documentary serving as a sad wake-up call to the situation in Agbogbloshie.

tinyurl.com/yyncao2g

Health issues and environmental problems accompany the issue of e-waste. E-waste contains numerous toxic substances that would call for safe handling and proper recycling. In European countries, companies such as Aurubis specialize in e-waste recycling (see pages 112ff.). In dumpsites, such as Agbogbloshie, in contrast, dangerous material fractions are burned, dumped or discharged into the environment. Dismantling happens without gloves or other protection. Experts say the problem is that people depend on these practices as a source of income. To close a site or prohibit practices would lead them to flee to other places or to increased illicit work.

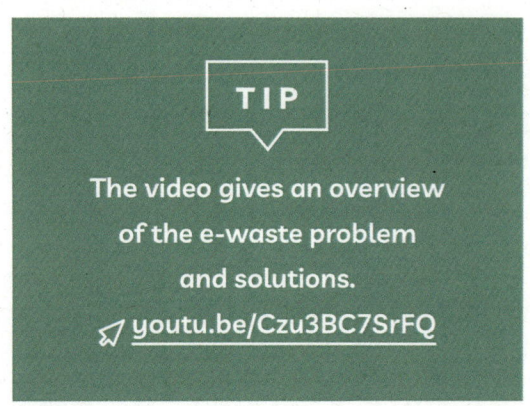

TIP

The video gives an overview of the e-waste problem and solutions.

youtu.be/Czu3BC7SrFQ

The circular economy is still a relatively new concept to e-waste, and also with regard to smartphones. Among the approaches to tackle the problem is a take-back program of phones from Ghana. Fairphone collected 75,000 phones in Ghana with the help of local collectors to recycle them properly in equipped factories in Europe.[177] Another approach is to build up a modern

THE CIRCULAR ECONOMY

The circular economy is often seen overall—and not just specifically with smartphones—as a major approach to tackling global waste challenges. It *"strives to minimize negative environmental impacts through qualitative transformation coupled with the closure and deceleration of material cycles. Circular economy approaches can take effect in the various stages of a product's lifecycle: Material selection and design should enable longevity, recycling and reparability or biodegradability. The service life should be optimized and prolonged. … At the end of the service life, the various materials should—as far as possible—be disassembled, sorted and recycled for further use."*[178]

The nonprofit institute Cradle to Cradle inspired the circular economy and can be considered as a specific approach. It targets a circular economy from the start of a product (cradle) to its end (cradle), focusing on product design that completely avoids waste: *"The Cradle to Cradle® design concept describes the safe and potentially infinite circulation of materials and nutrients in cycles. All constituents are chemically harmless and recyclable. Waste as we know it today and which is generated according to the pre-existing take-make-waste model will no longer exist, only useful nutrients."*[178] The first products from the textile, basic materials, building materials, home and office supply, interior design and furniture industries were already invented and re-designed along with these criteria.

Circular economy and cradle-to-cradle approaches are changing the way of thinking and planning business to close loops.

However, the World Resource Institute (WRI) also points out that the circular economy has to carefully evaluate the costs and benefits of recycling, and also the distribution of burdens. They warn: *"Recycling is only worthwhile from a resource perspective if the resources required for recovery and recycling are less than those required for extraction and disposal. Some have even suggested that the circular economy could result in resource depletion if not carefully implemented."* They recommend increasing research and impact-based metrics to identify the optimum recycling rate for each material. Additionally, the WRI mentions that shifting from virgin extraction to repair in the European Union would add 700,000 jobs, mainly in Central and Eastern Europe. At the same time, less virgin material extraction would imply fewer mining jobs.[179]

Three links with more about cradle to cradle.

✈ tinyurl.com/y555uxkn   ✈ www.c2ccertified.org

✈ tinyurl.com/y2ard8o4

recycling facility within the Agbogbloshie site, supported by the German Development Agency, including apprenticeships. In Nigeria, an initiative of the government, international organizations, the private sector and civil society, strives to develop systems for the disposal of non-usable and toxic waste. A key goal of the initiative is to collect, treat and dispose of more than 270 tons of e-waste contaminated with persistent organic pollutants and 30 tons of waste containing mercury.[175]

TRANSFORM FISHING—AND MEAT?

We saw that overfishing is the biggest stress on biodiversity when it comes to our oceans. Obviously, fishing is an industry that highly and very directly depends on natural resources. On the one hand, the numbers show extreme pressure on natural resources—such as fish diversity, overall populations, and the health of oceans. At the same time, consumption and the industry have dramatically grown over the past decades. Is this path at all sustainable?

A BOOMING INDUSTRY

20.5 kilograms of fish per year—that's the amount each person on Earth eats on average.[147] On a per capita basis, it's twice as much as consumed 50 years ago.

In 2018, global fish production reached 178.5 million tons—the highest level ever recorded. Production from both capture fisheries (96.4 million tons) and aquaculture (82.1 million tons) also achieved their highest levels. Since 1990, global capture fishery production rose by 14%, global aquaculture production by 527%. By 2030, total fish production is expected to rise further to 204 million tons, as forecast in the 2020 edition of The State of World Fisheries and Aquaculture by the FAO.[147]

The top seven countries for global capture fisheries production accounted for almost 50% of total captures, led by China (15%) and followed by Indonesia, Peru, India, the Russian Federation, the US and Vietnam. The top 20 producing countries accounted for about 74% of the total capture fisheries production.[147]

World aquaculture production of farmed aquatic animals is dominated by Asia, with 89%. Major aquaculture producing countries include China, Bangladesh, Chile, Egypt, India, Indonesia, Norway and Vietnam.[147]

Fish and fish products are some of the most traded food items in the world. In 2018, 67 million tons of fish (live weight equivalent) was traded internationally for a total export value of 164 billion US dollars.[147] The European Union is the world's primary importer of fish.[180]

TIP

Sustainable fishing practices are just beginning to spread to the mass market. A link listing some initiatives in this area.
⇗ tinyurl.com/y5ju3c3g

HOW ARE FISHERMEN DOING?

In the fisheries and aquaculture sectors, an estimated 59.5 million people were directly employed: 39 million in fisheries and 20.5 million people in aquaculture (2018). 85% work in Asia and 9% in Africa.[147] According to the United Nations, marine fisheries directly or indirectly employ even more than 200 million people worldwide, and the livelihoods of more than 3 billion people depend on marine and coastal biodiversity.

Perspectives for fishermen change, such as for traditional artisanal fishermen in Peru´s capital Lima. Their business is endangered by increased competitive pressure and gentrification—meaning cities increasingly become attractive and thus, change their structure as financial interests push prices higher.[181]

TIP

A fish directory that shows consumers which fish came from which countries and what labels are recommended. It is frequently updated by the WWF and can be downloaded as an app in different languages.
⇗ tinyurl.com/y6euvo96

Working conditions in the fishing industry have not been in the public focus much until now. An analysis from the early 2000s described precarious conditions for small-scale capture fisheries: *"in highly populated Asian countries, artisanal fishing families are among the most socially, economically and politically disadvantaged segments of the population and maintain a status comparable to that of landless* *laborers or marginal farmers. Deprivation is so severe that the basic needs of life are hardly met at the minimum level necessary for survival. Malnutrition is common, infant mortality is high, and chronic sickness and disease result in very low life expectancies. Conditions are similar in several areas of Africa and Latin America. However, small-scale fishing families are generally better off on these continents, even if the average income levels in small-scale fisheries are often below the official poverty lines. According to FAO estimates, the number of poor small-scale fishers and related employees in marine and inland capture fisheries is 5.8 million, representing 20% of the world's 29 million fishers, and they earn less than US$1 a day. There may be as many as 17.3 million income-poor people in related upstream and downstream activities, e.g. boat-building, marketing and processing. These figures suggest an overall estimate of 23 million income-poor people, plus their household dependents, who rely on small-scale fisheries for their livelihoods.*"[182]

The FAO report from 2020 mentions cases of appalling working conditions within the capture fisheries sector, especially related to illegal, unreported and unregulated fishing (IUU): *"Abuses have been reported in fish processing plants and on board fishing vessels, where working conditions are more difficult to monitor. There are strong indications that human trafficking, forced labor and other labor abuses on board fishing vessels are associated with IUU fishing, with migrant workers identified as a particularly vulnerable group.*"[147]

WHAT ABOUT OVERFISHING?

The issue we are aware of is the ecological challenge related to overfishing. The FAO reports that the percentage of fish stocks within biologically sustainable levels decreased from 90% (1990) to 65.8% (2017).[147]

In 2017, the areas among the FAO's 16 Major Fishing Areas that had the highest percentage of stocks fished at unsustainable levels were the Mediterranean and the Black Sea (62.5%), followed by the Southeast Pacific (54.5%) and Southwest Atlantic (53.3%). The Eastern Central Pacific, Southwest Pacific, Northeast Pacific, and Western Central Pacific had the lowest proportion (13–22%) of stocks fished at biologically unsustainable levels.[147]

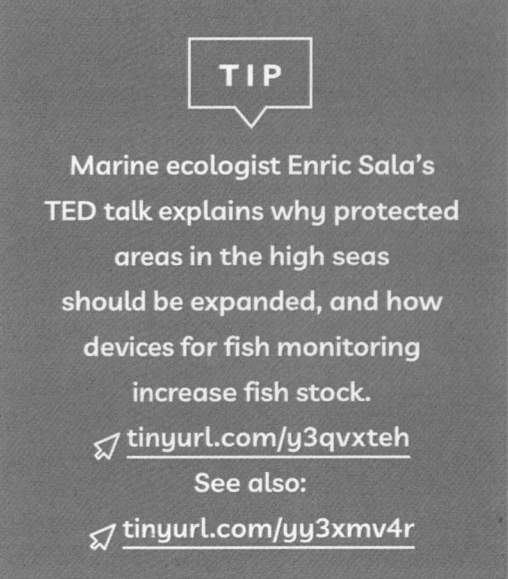

TIP

Marine ecologist Enric Sala's TED talk explains why protected areas in the high seas should be expanded, and how devices for fish monitoring increase fish stock.

🖅 tinyurl.com/y3qvxteh

See also:

🖅 tinyurl.com/yy3xmv4r

These figures, however, can be misleading. Of the 600 marine fish stocks monitored by the FAO, more than 70% of the world's fish species are fully exploited or depleted.[183] Over just 40 years there has been a decrease recorded in marine species of 39%.[180] Yet, the United Nations also reported that global marine key biodiversity areas covered by protected areas rose from 30.5% (2000) to 46% in 2019.[105]

One major challenge in overfishing is bycatch, resulting from the currently preferred fishing practices. It accounts for about 30% of bycatch each year, equaling about 38.5 million tons.[180] The FAO estimates that as high as 35% of the global harvest of fisheries and aquaculture production is either lost or wasted.[147]

Trawling, for example, is critical due to high amounts of bycatch. It is a practice affecting 19 million tons of fish and invertebrates annually, or about 25% of world's fishing.[185] For one kilo of shrimps, up to 9 kilos of bycatch result, as shrimp trawlers throw away up to 80 to 90% of bycatch.[185] Studies showed that the environmental effect of heavily trawled areas was most significant.[185] Only in areas such as Alaska, trawling causes less damage: This is mainly due to an overall comparably small amount that is trawled, small areas are fished and fixed quota for bycatch exist. Also, economic pressure for fishermen is lower than in other areas of the world.[186]

A project targeted to reduce bycatch in Latin America and the Caribbean included sharing best practices in bycatch reduction technologies and improving institutional and regulatory frameworks. In some cases, reductions of up to 30% of bycatch were achieved.[187]

Another challenge, again, is the illegal, unreported and unregulated fishing (IUU) that is estimated to account for 11-26 million tons (12-28%) of fishing worldwide.[180] For several years, combating illegal, unreported and unregulated fishing has been a priority on the agenda of international negotiations and intergovernmental organizations.[147] A driver was that IUU became part of the Sustainable Development Goals, specifically it is targeted within SDG 14.4 and 14.6 to end IUU fishing and eliminate subsidies that contribute to IUU fishing. According to the UN, 97 countries to date have signed the Agreement on Port State Measures, which is the first binding international agreement on illegal, unreported and unregulated fishing.[105]

TIP

A podcast explaining how COVID-19 has disrupted fisheries around the world, and how conservation efforts can come back from the pandemic even stronger than before.

tinyurl.com/y6aqqwc8

GARDENING FISH

Can the fast-growing aquaculture business provide a solution? This method of artificial fish farming is taking the stress out of natural resources. The advocacy group Aquaculture Alliance defines it as: *"Aquaculture is the controlled process of cultivating aquatic organisms, especially for human consumption. It's a similar concept to agriculture, but with fish instead of plants or livestock. Aquaculture is also referred to as fish farming."*[188]

Since the beginning, aquaculture has faced skepticism as a result of several issues, mostly environmental but also related to animal welfare. It is encouraging that innovation has improved the industry's balance in recent years, but still, the industry is on the way to becoming sustainable. A key success factor will be not to overstretch its positive contribution but to discuss openly and in a balanced way, the challenges of aquaculture to avoid greenwashing.

A major challenge is feeding, as most aquaculture does need natural fish. Some argue that when considering it as an aggregate industry, global aquaculture uses about half a metric ton of whole wild fish to produce one metric ton of farmed seafood. And that farmed fish is far more efficient at converting feed than wild fish or other farmed animals such as cows and pigs.[189]

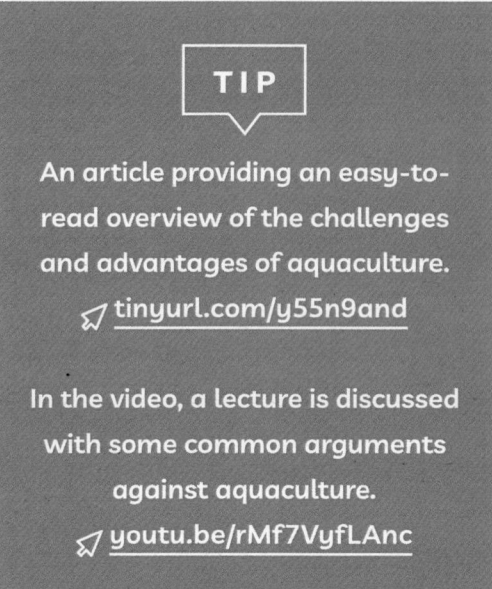

TIP

An article providing an easy-to-read overview of the challenges and advantages of aquaculture.
⤤ tinyurl.com/y55n9and

In the video, a lecture is discussed with some common arguments against aquaculture.
⤤ youtu.be/rMf7VyfLAnc

Still, some species consume much more fish by weight then they eventually produce, among them shrimp and salmon. For example, for one kilo of aquaculture salmon, up to 5 kilograms of other fish are needed to feed the aquaculture. This means that through aquaculture, new fish are not created but, according to Marinebio.org "less valuable fish" are converted into "more valuable fish".
⤤ youtu.be/Q1EcQQsJrmg

Cargill, one of the largest global suppliers of aquafeed and a leading supplier of salmon feed, conducts research to increase the efficiency of feed production and conversion. They state: *"In our salmon feed business, for example, we have decreased our use of marine raw materials (sum of fish meal and fish oil) by half, from 55% in 2005 to 27% in 2015, despite a large increase in annual feed production over the same period."*[190]

Within the industry, some economic mechanisms trigger improvements: Feed for farmed animals is the most expensive input for fish farming; it accounts for about 60% of the cost of growing fish. Over the past 20 years, fishmeal—one of the major products for feeding—has seen a 110% price increase and has been largely volatile. This implies that, while in 2000, over 10 ounces of wild fish were caught for every pound of farmed seafood, in 2015, less than 4 ounces of wild fish were caught. Environmental considerations are thus accompanied by economic incentives for fish farmers to avoid over-feeding and a search for alternatives. Farmers use advanced technology, including automated feeders and underwater cameras, to provide feed and monitor consumption. Alternatives to feeding natural fish are frequently searched for. They include advanced plant-based products, algae and insect protein.[191]

Other challenges from this type of farming in controlled conditions include, among others, pollution to discharged water and releasing chemicals and antibiotics into nature. The high density of the fish population can cause diseases. Destruction of coastal ecosystems, such as mangroves for shrimp farming, was a major issue in the early days of aquafarming. According to the Global Aquaculture Alliance, this has improved in the past several years. Aquaculture also leaves a footprint as a result of the use of freshwater and energy-intensive heating and cooling, as well as from the use of pump systems. Animal welfare is another topic, including the questions of whether raising aquaculture in a tank is appropriate and how breeding should be managed.

TIP

Websites that show which products at which retailers in countries around the world have labelled fish.

Aquaculture (ASC label)
tinyurl.com/y49pqg26

Other fish (MSC label)
tinyurl.com/yygo27fn

Several activities to improve aquaculture are on their way at different levels. The FAO is currently developing Sustainable Aquaculture Guidelines (SAG), mainly targeting policymakers.[147] Based upon that efforts, Sustainable Fisheries Partnership (SFP) promotes a Framework for Sustainably Managed Aquaculture across the supply chain and engages, for example, in stakeholder roundtables. Aquaculture without Frontiers (AwF) focuses on small-scale sustainable aquaculture that alleviates poverty, integrated into rural and coastal development plans. Technology transfer and capacity building with farmers, for example, in Malawi, Cambodia and Mexico is conducted mainly by volunteers of AwF.

The long-recognized label Marine Stewardship Council (MSC) certifies environmentally and socially responsible seafood in the capture fishery. Ten years ago, the Aquaculture Stewardship Council (ASC) was established for aquaculture. The ASC publishes a monthly update on its coverage. In July 2020, for example, almost 25,000 products in 90 countries were labeled, and 49% of all shrimp aquaculture was labeled but only 3% of Talapia.[192] This website also makes it possible to search for products by country.

In summary, fishing involves interesting dynamics at several levels in an effort to find a balance between economic, environmental and social interests—as well as between people's needs and the ecological balance in our oceans.

HOW ABOUT MEAT, THEN?

Aquaculture is accepted and appreciated by consumers, as the growth rates show. Innovations are moving from natural to more artificial farming. They create desired outputs, such as certain species and flavors, lower footprints and high conversion rates.

The trend towards more artificial farming is slowly emerging in the meat industry. This will have a significant impact on nature and could disrupt an established industry. The global meat market is huge, about 1,000 billion US dollars in size (2018), according to a recent study by A.T. Kearney. Its footprint is huge, too: Animal farming uses about one-third of global land, including cattle ranching and growing feed. Nearly half of the worldwide harvest is required to feed the livestock population.[193] Out of the food sector's carbon footprint, which accounts for 26% of global carbon emissions, the footprint of livestock farming takes about a third.[194]

According to the A.T. Kearney report, conventional meat production has reached its limits to meet the growing global demand, especially in terms of the availability of land and water. They also argue: *"All predominant innovations, including digitalization, simply increase effi-*

> **TIP**
>
> A worthwhile report by A.T. Kearney, "How will cultured meat and meat alternatives disrupt the Agricultural and Food Industry?" explains, in detail, the methods of cultured meat and which companies are active in this area.
>
> tinyurl.com/y2gl9k6e

ciency of conventional production methods and won't overcome global agricultural and food challenges."[193] Meat replacement has the potential to disrupt the industry and address these challenges, they say. Thus, it is a logical consequence to search for alternative scenarios.

Some years ago, producing meat artificially was almost unthinkable for consumers. But with growing trends such as vegetarian and vegan meat alternatives, transformation is accelerating.

A.T. Kearny predicts that more than a third of our future meat consumption, such as steaks or burger patties, could be so-called cultured meat (for example, chemically manufactured or even using even 3D technology) by 2040. Companies like Mission Barns, Aleph Farms and IntegriCulture work intensively on these types of manufactured meat. Venture capital is increasingly flowing into the sector. Meanwhile, sample products are available but not yet on the market or have passed regulations. Currently, production methods are not yet feasible from an economic perspective and are still very energy-intensive. Both are expected to improve with higher scalability.

It will be interesting to see how this trend evolves, and how we will behave as customers. Many ethical issues and sustainability challenges related to conventional livestock farming could be overcome through methods similar to aquafarming. Will we accept functional food in meat to alleviate pressure on nature and people? Consumer acceptance will be a key decision point, as it is in many other areas, too.

APPLICATION

1. ———— Reflecting on the different industries and their business dynamics (tourism, smartphone, textile, fishing, meat).
   – What challenged or moved you? Why?
   – What solutions can you think of for the displayed challenges?

2. ———— For your company context:
   – What are the dynamics that may need to be changed?
   – What could be the likely approaches?

3. ———— Take a quiz!
   Earth Day offers several quizzes, for example on climate, environmental literacy and clean energy.
   ⟡ www.earthday.org/earth-day-quizzes

4. ———— Conduct an "Urban Mining Workshop" with friends or colleagues to explore the story behind your smartphone.
The material for this is provided open-source by Fairphone. ✈ tinyurl.com/y3f46vsy

5. ———— Discover with kids!
Kids Environmental Lesson Plans (KELP) are free, downloadable activities that educate children about marine science. They offer games, handicrafts and other activities with minimal preparation, for topics such as ocean health issues including plastic pollution, overfishing and climate change.
✈ tinyurl.com/y34nybug
✈ tinyurl.com/y6sdkcw9

ABOUT YOU AND ME

"I am Greta" said a relative recently, pointing to a few good deeds like recycling, avoiding plastic and frequent bicycle use. After attending a workshop on more sustainable lifestyles, one friend questioned whether she should now make her own soap and grow her own vegetables for her family of five. Another friend returned a small Christmas gift I had sent at the last minute and wrote me a long e-mail explaining it as an act of rebellion against a huge parcel delivery company—from both a sustainability standpoint as well as a global income disparity perspective.

When one works in sustainability, it naturally interferes with daily life. I try to live by high standards based on what I find important, but I am not a sustainability saint. I might know a bit more about sustainable choices and make a few more decisions on that basis, but, in the end, it is an ongoing process of making wiser choices—for you and me.

WHY MORE SUSTAIN-ABILITY IS NOT ALWAYS GOOD

If you read through magazines, blogs, book releases, podcast lists and traditional or social media, sustainability is now glaringly covered. You get frequent reminders and ten-point lists on how to live and shop greener— sometimes with an underlying hint of pressure. We read how electric vehicles, sustainable food and green fashion trends have skyrocketed. Retailers, hotels, shops and online marketplaces promise you that they are taking sustainability into consideration. People are standing up against climate change at Friday's for Future demonstrations. Across cities, you see urban gardening, car- and bike-sharing and vegan cafés mushrooming. This could give us the illusion that sustainability has arrived in our lives and that there are many people highly engaged and living more sustainable lifestyles. We are likely to think that, although the situation is severe, we are at least doing something about it and somehow seem to be on the right track. As we embark on our own personal sustainability journeys, I would like to share a few observations of why I have a critical view of this current narrative, even to the point where I believe it is often counterproductive.

IT DELUDES THE FACTS

It is a simple economic logic that market share grows over time and that growth rates tend to be high in the early stages. Sustainable choices are on the rise—but in niches. We should celebrate growth because it shows that change is possible and motivates people. We must, however, be careful that the common public rhetoric doesn´t trick us into thinking that there has been more positive progress than really exists or delude us from seeing the real nature of the status quo and the continuing challenges. There is a danger of greenwashing—not by corporations but by public opinion. Even though this is not likely the intention, the danger of delusion still exists.

The following are two examples that illustrate why context and complete information are relevant and how rhetoric that is too positive can distract us from the realities.

"It's 2020 and electric cars are paving the way to a healthier planet", was stated in a recent article posted online.[195] Don't most of us want a healthier planet? Solutions like electric cars that makes the world better? Answers that take some of the pressure off about climate change? Things that take away the guilty feeling we get sometimes when we drive our own car? It's a good feeling to know that things are being taken care of.

The facts: In 2019, 5.6 million electric vehicles were on the road worldwide. This figure includes passenger cars as well as any commercial vehicles you may have seen, such as parcel delivery trucks. According to the article, this number is larger than ever before and had increased by an impressive 64% since 2018.[195] In 2019, that same year, 65.5 million "regular" passenger cars were sold. That means that more than ten times the number of all existing electric vehicles were added to our roads worldwide.[196] That doesn't sound very much like we are getting closer to a healthier planet or give us any reason to think some of the pressure is off about climate change when it comes to mobility.

When searching the web and social media for green fashion, tons of results pop up that also give us the feeling that things are going in the right direction. Style magazine "Elle" made this statement already in 2014: *"But as the eco-friendly fashion market continues to grow, new crops of labels are taking the industry by storm ... see why shopping has never been so guilt-free."*[197] Isn't that what we want to hear? That fashion is becoming increasingly eco-friendly, and new "good" labels are successful? No more feeling guilty? In fact, our shopping is even making the world a better place?

The facts: Yes, there has been some market growth, multiple new labels and more awareness. Still, the fashion industry is far from being greener, more sustainable or fairer, as we have already extensively seen. The trend towards fast fashion predominates—leaving a huge social and eco-footprint. Rhetoric on eco-fashion and sustainable apparel leads us to believe differently.

As sustainability is complex, it's often a challenge to break it down, contextualize well, and make it practical and motivational. Yet, it will be a continuous task for all of us to avoid and prevent vague or biased information that nurtures an illusion of a common understanding or progress that is non-existent.

IT LEAVES US COMPLACENT

Deluded facts can leave us complacent as it can result in the feeling of "It's all taken care of". There is however another trend that may lead us to become too contented: Sustainability has entered our daily lives as a lifestyle topic.

Rhetoric nurtures this profoundly and repeatedly. Niche cosmetic products stress their "saving the world" characteristics. Wellness hotels claim to be sustainable to appease our yearning for feeling good—whether these hotels are really focused in the end on the right things has, so far, not really been questioned.

TIP

Three Instagram feeds
with good tips.
@sustainable.collective
@get.waste.ed
@onesaveaday

In magazines, books and blogs, you can find countless tips on urban gardening, self-made gifts, recycling, upcycling, vegan eating, living without plastic, biodegradable goods, green cosmetics, self-made soaps, sharing (cars, homes, appliances, clothes, meals, time, etc.) and so on. There are many people busy doing good things. Self-optimization tips and instructions in lifestyle magazines and blogs on how to live and shop greener give us guidance. Doing good makes people feel good.

While it is commendable—and necessary—to take steps in this direction, as will be shown later, there is a danger in using this lifestyle approach. On the one hand, it has a tendency to become simplified and superficial. Substance, usefulness, relevance, and authenticity are not always questioned. Sustainability is treated as a trend, and thus claimed and labeled wherever possible—precisely the same approach that upsets people when companies greenwash (see chapter 5). Overuse in this area is already evident, which deteriorates credibility and prevents the necessary actions from being taken.

Especially when we just do what's trendy, fun, or convenient—as is often the case with the lifestyle approach—the changes we make will be solely based on our personal preferences but not really on necessity. This tendency nurtures our prevalent resistant behavior, which we will analyze shortly. By using the lifestyle approach, we are also probably overestimating our efforts and contributions. People are kept busy doing things that, even though they may have little effect, give them a good feeling and make them complacent. Complacency prevents well-informed, mature choices that would be taken freely and in the interest of relevant action. Sustainability requires a continuous way of evaluating and making wiser choices— small and big ones, easy and challenging ones, in different areas of life.

IT CREATES GUILT AND PRESSURE

Another approach that has developed is to bash, ban, overcriticize and exert pressure, in a subtle or more direct way. Our lives are already filled with judgment, and increasingly so when it comes to sustainability. Sometimes, our current mainstream narrative puts pressure on people to behave in a certain way without true conviction, motivation or, sometimes, even the ability to act.

While it is necessary to learn the facts, and motivate and challenge people, we need to walk a very fine line. For example, it is good and necessary to cut down on flying and consider alternatives whenever possible—but a term like "flight shaming" overreaches. This term is used in articles and campaigns and on social media to bash people who use airplanes and to make people feel guilty about flying. Finger-pointing also occurs frequently when it comes to eating choices (vegetarian, vegan vs. meat eaters) and, recently, with regard to video streaming. I agree with other experts that certain options, such as flying, are avoidably unsustainable and that some choices, such as vegetarianism and recycling, are more sustainable than others. But we live in democratic countries where freedom and individual decisions are, thankfully, highly valued. Making people feel ashamed or guilty is, from my perspective, neither a respectful nor useful approach. Profound change will not happen if people act only out of fear or shame.

Public discourse about better solutions is necessary—ideally in a clear, respectful and solution-oriented manner. It is useful to publicly discuss what is appropriate, especially when role models are involved, as they have a big influence on many people. When the German national soccer team took a plane in summer of 2020 for the 260 km distance from Stuttgart to Basel, they encountered immense criticism on social media. Taking a train or bus would have been a better solution for the environment. The officials argued that the performance and health of the soccer players were being prioritized. Supposedly the criticism triggered some to take a stronger look at and equally consider environmental effects, it was stated. That remains to be seen, but it is a promising sign of progress when German soccer is publicly criticized in favor of the climate. It would be great to see role models begin to take on more responsibility for sustainability topics.

To sum it up, when it comes to a good cause, we are sometimes tempted to point out what others are doing wrong and what they should do—I explicitly include myself here. This puts pressure on people to behave "right", as was the case with my friend who felt socially obliged to try urban gardening. Instead, we should strive to empower people to make wiser decisions in consideration of the consequences of consumption on people and nature. This would require information and action in the relevant areas of consumption and will sometimes require confrontation, also with respect to underlying motives. But we should let our passion for change be led with respect and in the spirit of freedom.

WHAT'S GREAT ABOUT OUR CURRENT NARRATIVE

Aside from the three points mentioned where I am critical of our narrative, it is great that sustainability has leapt forward in terms of public awareness. Let's, therefore, honor some of the good aspects of the narrative.

There is much more consciousness today than in the past, now that we are dealing with serious topics such as climate change, resource scarcity, waste on land and in oceans, endangered species, animal welfare, health problems, educational divide, unfair wages and labor conditions, poverty and so on. These are no longer topics addressed only by experts and researchers. They've come closer.

The awareness is increasing that these issues involve each and every one of us. The realization that our consumption patterns are not sustainable and our lifestyles harm nature and people is also sinking in. We are becoming more sensitive to patterns that are detrimental.

Tremendous care and the best of intentions go into all of the approaches and their communication. We are not looking to destroy nature. We are not indifferent to animal welfare. And we do not want to be responsible for people working under precarious conditions. Children working to produce for our consumption is socially unacceptable. It seems like a new social consensus is starting to emerge that sees sustainability as the right direction.

There is empowerment and mobilization. People are standing up and engaging. They are participating in Friday's for Future gatherings and advocating for better working conditions in the healthcare sector amid the pandemic. People are trying out new things and intending to make a difference in their day-to-day lives. Concepts such as urban gardening and upcycling are spurring creativity. Broad segments of the population are taking more care in their direct surroundings, trying to reduce plastic, eat less meat or become vegetarian, and searching for ways to live and shop more consciously. Publications, magazines, and blogs are informative and motivational, and encourage people to do good. There is much more information available today than ever before.

The enormous amount of creativity, heart, good intentions, concrete actions and impacts triggered through a sustainability movement across the population is encouraging. Yet, it still has not become a mainstream topic with "everybody on board".

10.02 ———— YES, THE CONSUMER MATTERS—A LOT

While corporations are tremendously responsible and crucial for advancing sustainability, consumers matter, too. They play a major role as their decisions and demands guide companies. They can act as powerful enablers. Though we have seen some progression, the potential of consumers has not yet been unleashed. A realistic view is necessary in order to forge ahead. Let's look at some facts, relevant interrelations and hurdles.

WHY DIFFERENT MAKES A DIFFERENCE

What are the consumption areas most relevant for reducing carbon emissions? In Germany, the CO_2 contributions from private consumption break down into living (39%), mobility (26%) and food (15%), with a further 13% attributed to the consumption of goods (appliances, textiles, hygiene products, paper, furniture) and other consumption.[198]

The analysis suggests that German consumers, among other things, should increasingly switch to green energy, increase environmentally friendly heat generation, install energy-efficient household appliances and lighting, buy more organic products and labeled fish, use more public transportation and car-sharing and switch to electric and hybrid vehicles when driving.

Let's dive into the topic of food—definitely the most longstanding, advanced area in Germany.

– The share of organic food has been growing across the world in recent years. Germany is one of the leading countries in per capita spending on organic food, just below Switzerland, Denmark and Sweden. The market share of organic food in Germany, however, is just 5.7%.[199]

– Fairtrade products pay farmers a bonus in response to fair wages, which is a social topic prevalent for many agricultural products. The total revenue generated with Fairtrade products worldwide was 9.8 billion euros in 2018, of which 1.33 billion euros were attributed to Germany. Despite encouraging growth rates, this amounts to a market share of no more than 0.7%.[200]

– The top-selling sustainable food segment in Germany in 2019 was regional products, which accounted for 17.9% of market share.[200] Regionality fosters local value creation and limits transportation.

– Germany is the European country with the greatest selection of vegan products, the highest number of new product launches in the segment, as well as significant growth rates across supermarkets and other channels.[201] To put this in overall context, only 1 million people out of a population of 82 million in Germany are vegan, and another 6 million live as vegetarians.[202] These figures however do not reflect the increasing number of people incorporating vegan and vegetarian choices into their diets.

– Per capita meat consumption has decreased in Germany to 59.5 kg per year (2019) compared to 62.8 kg (2011).[203] Still, 28% of the population eats some sort of meat daily.[202] Meat is cheap—Germany ranks tenth in terms of prices across European countries.[204]

– 70% of the German population claims to care about animal welfare. Yet, 95% of meat production   and consumption—is conventional with a comparably low level of animal welfare.[205]

– Each year, 11 million tons of food is thrown away in Germany, amounting to more than 130 kilograms per capita.[206] A national strategy is targeting a reduction in food waste of 50% by 2030. ✈ tinyurl.com/y4qltdj8

– Packaging waste has grown dramatically in Germany and amounts to 18.7 million tons per year (2017), or 220 kg per capita, making it the country with the most packaging waste per capita in Europe. Plastic packaging material has doubled since 1995.[207] Less than a third of all plastic waste in Germany is recycled; the rest is exported or combusted. One reason for the amount of waste is the lack of recyclability of many packaging materials and low recycling rates. A new law enforced on packaging set out a goal to increase the recycling of plastic to 63% by 2022.

In sum: As most sustainable choices are in niches, choosing wisely can make a difference in each single supermarket visit. It also shows that other topics matter significantly, such as plastic use, recycling and food waste.

In the US, sustainable choices in supermarkets are also on the rise, but here too, this is occurring mostly in niche areas. A recent Harvard Business Review article showed that consumers in the US are spending an ever-growing amount of money on products that are labelled with a sustainability claim. The numbers varied across categories, with the highest share in the category (more than 18%) reached by toilet tissue, facial tissue, milk, yogurt, coffee, salty snacks, and bottled juices, while laundry care, floor cleaner, and chocolate candy had less than a 5% share.[208]

TIP

A video (available in English and Arab) raises awareness of the high footprint citizens from the UAE produce. It does provide just a few hints on how to actually reduce it.

tinyurl.com/y2ppx6q4

In the United Arab Emirates, one of the countries with the highest per capita carbon footprint, a study showed that a high 57% of the country's footprint was attributed to private consumption. A significant factor was the enforcement of energy-efficient light bulbs that have a significant impact on reducing energy use and, in turn, emissions at the household level in the UAE. These results show that improving the personal carbon footprint also requires enforcing political regulation.[209]

LESS IS ACTUALLY PART OF THE GAME

Switching to sustainable choices is great but it is not enough. We need to consume less. Consumption throughout the world is accompanied by high burdens. At a global level, we currently consume the equivalent of what would be consumed by 1.7 earths. The United States is the country with the highest per capita consumption, with consumption equivalent to that of 5 earths.[210] This takes the overall environmental footprint into

How many earths do we need if the world's population lived like:

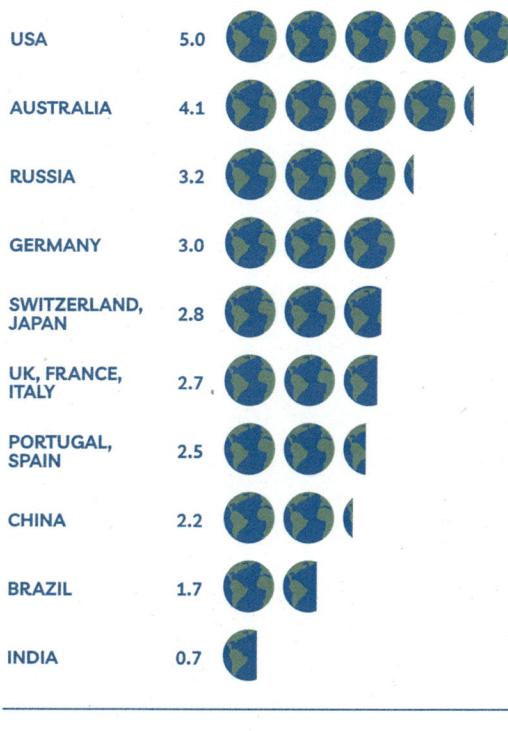

USA	5.0
AUSTRALIA	4.1
RUSSIA	3.2
GERMANY	3.0
SWITZERLAND, JAPAN	2.8
UK, FRANCE, ITALY	2.7
PORTUGAL, SPAIN	2.5
CHINA	2.2
BRAZIL	1.7
INDIA	0.7
WORLD	1.7

Own graph based on source: Global Footprint Network website. tinyurl.com/y2naap4e. Accessed December 15, 2020.

consideration, including emissions, land use, water and resources and is broader than the perspective for carbon only.

Each year, *"Earth Overshoot Day marks the date when humanity's demand for ecological resources and services in a given year exceeds what Earth can regenerate in that year."* In 2020, Earth Overshoot Day was again on August 22. A campaign was held as #movethedate. www.overshootday.org

Not surprisingly, it is people with higher incomes that create more than their share of the environmental burden. The average carbon footprint of the wealthiest households in the US is over five times that of the poorest households. However, this is not only a matter of the super-rich. US consumers with more than 100,000 US dollars in annual household income accounted for one-third of the household carbon footprint while representing 22% of the total population.[211]

This type of relationship is also true for waste, which is another major topic in the world today. Each person in the world produces 0.74 kilograms of waste on average every day. This number rises significantly the richer people get. This figure is more than 2 kilograms per day in rich countries like the US, Switzerland, Denmark, Hongkong, and New Zealand.[212]

Consumption patterns vary, certainly across countries and products. The data for a simple product like toilet paper reveal an astonishing range.

Per capita consumption of toilet paper (per year, 2018)

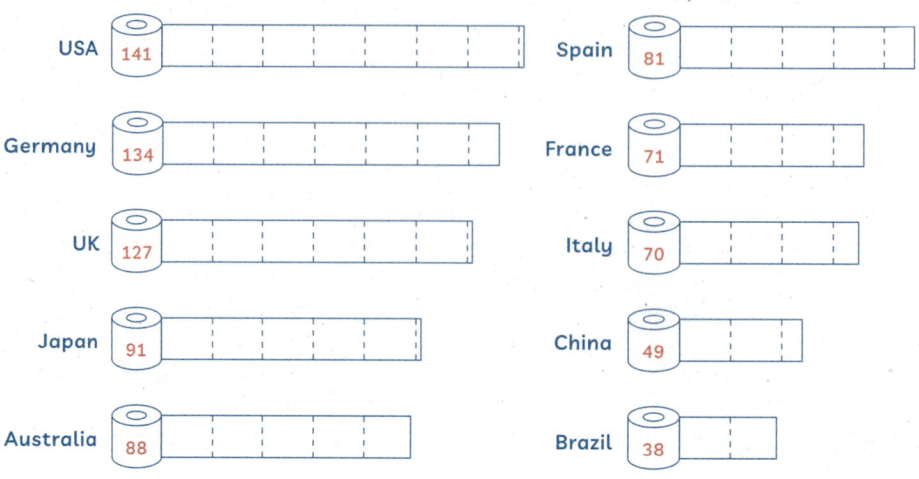

USA 141
Germany 134
UK 127
Japan 91
Australia 88
Spain 81
France 71
Italy 70
China 49
Brazil 38

Own graph based on source: Statista website. https://tinyurl.com/yx2aukjc. Accessed October 28, 2020.

The overall conclusion, which applies not only to and within rich countries, but also to higher-income populations is simple: We need to consume less. Let's turn to the US example as it is significant in terms of consumption and emissions.

TIP

A tool that helps US consumers easily measure their carbon emissions from housing and detect the potential for reductions. tinyurl.com/h9l5ygo It is provided by the Environmental Protection Agency (EPA) and specifies the national context.

Food, which is one of the categories in supermarkets where sustainable choices are on the rise, contributes 16% to the domestic carbon budget of households, and is therefore comparable to Germany on a relative basis.

The most significant areas of consumption leading to carbon emissions in the US are housing and transportation, accounting for over 60% to the total domestic carbon footprint of US households. What this means is that if consumers in the US want to reduce their carbon footprint, they should start with housing, which accounts for 25% of energy use.

TIP

A fact-filled article, "The five elephants in the room", states that electricity, natural gas, gasoline, flying and food are the dominant leverage points in reducing the average American's footprint. More than 60% of the footprint originates from these five areas, even though they account for only 15% of spending.

tinyurl.com/y43v3ebm

Additionally, US consumers would need to change their transportation habit (see page 329). Since 1995, the household carbon footprint from transportation has risen significantly. This includes fuel but also emissions from vehicle manufacturing. Transportation emissions per household and per capita have continued to rise despite a significant reduction in exhaust emissions from vehicles and a nearly 30% improvement in the fuel economy of cars between 1995 and 2014. Why? Because people in the US travel more by car, own more household vehicles, and travel their vehicles with a lower average number of occupants.[211]

To consume less and less carbon-intensively is a significant step in combating climate change. It also helps to reduce resource exploitation as well as the use of natural resources. The effect are not as evident when you look at the negative social dynamics in supply chains: less consumption does not automatically improve the conditions and income of workers in poorer countries. On the contrary. This is why the combination of less and different is so relevant.

HOW DO WE USE OUR POWER?

The underlying argument in this book is that our consumption has an influence. As consumers, we have the power to change the course of things or make a difference by how we act, even though we often don't think we do. It might seem right or convenient to transfer responsibility to others for big topics such as climate change, animal welfare or global income disparity. But the interrelation may become clearer and more direct when taking a closer look. Let me illustrate this with the example of global income disparity, a huge and emotional topic for many people when you look at the facts.

"The richest 1% own 44% of the world's wealth", states the Credit Suisse Global Wealth Report.[213]

"Here's how the ultrawealthy got even richer during the pandemic while millions of Americans faced job loss, hunger, and homelessness" titled an article, stating that between March and September 2020, *"American billionaires have become 29% richer than they were before the pandemic ... Jeff Bezos profited so much that in August he briefly became the first person in history with a personal net worth over $200 billion."*[214]

Who is responsible for global income disparity? There are many answers to this; private consumption is part of it. Let's stick with the example of Amazon, where this becomes blatantly evident. *"Amazon: The Consumer's Everything In A Post Covid World"*—titled business magazine Forbes, stating that consumers spend an average of 40% of their total digital wallet with Amazon.[215] It is not a secret that Jeff Bezos owns shares in the company he founded and that he has been the world's richest person since 2017. Each purchase on Amazon's platform contributes to the company's success and make Jeff Bezos ultimately richer. Decisions have an effect.

> **TIP**
>
> The article looks at Amazon's activities during Covid-19.
> ↗ tinyurl.com/yakd4fyy

It's not about a specific person or company. This example is meant to illustrate a point we often gloss over as consumers: It is our responsibility to act on things that are important to us—not the responsibility of companies, politicians, or others. Too often, consumers argue that their decisions do not have an impact or that they cannot change the way things are. I frequently hear statements like:

"The airplane is going to fly anyway, regardless of whether I book the flight."

"If it's that cheap, I can just buy it/buy a second item."

"If companies are willing to offer free returns for online purchases, I can order a second item and just send one back."

Factually, there is nothing wrong with these arguments. Morally, I would not like to judge them. Each of us, luckily, has the freedom to make our own decisions. But we are reminded to consider the consequences and that we can make a difference when things are important to us.

If global income disparity (or the disappearance of small shops in city centers) is something I feel strongly about, then online consumption is simply not a smart decision. If I care about emissions, I can choose not to order two items just to send one back. If I want airlines to cut down on their services and reduce their climate impact, I should avoid taking an airplane as much as possible. If I think the parcel delivery guy has a tough job and earns too little money, I can simply choose not to order online, or give him a tip.

I encourage us to recognize our responsibility and the power we have as individual consumers. With each purchasing decision we make and each cent we spend, we make a difference anyway. It just depends on which direction we choose to exercise our power, who stands to benefit, and who assumes the burden. Consumption is the most powerful daily decision we make. That's why I am so passionate about empowering people to make wiser decisions. It can also lead to a clearer perspective about personal motives and priorities. Do I really care or is my outcry against global income disparity half-hearted? What am I willing to sacrifice for things that matter to me?

10.03 ——— WHY IS IT SO HARD TO CHANGE OUR CONSUMPTION?

We saw in the last passages that consumers play a significant role based on their consumption decisions. They influence market share, as well as other big issues like global income disparity. Consumers would need to change and reduce their consumption. But what's stopping us—what is standing in the way of you and I embarking on a more sustainable path? Why do we frequently make unwise decisions?

GAPS IN KNOWLEDGE AND ATTITUDE

For many years, experts have been trying to understand what is going on in consumer's minds and what deters sustainable behavior. Market researchers have been puzzled for years by the fact that consumers claim that they prefer sustainable products and intend to purchase them but their behavior in supermarkets doesn't reflect this. Consumers say they want to see better animal welfare, but still buy conventional products. Morally, they don't tolerate poor working conditions in the supply chain—but they also don't look at the available data and make decisions accordingly. Most would probably not want to contribute to climate change, even though they do so anyway through their manner of consumption.

Surveys also frequently report a readiness of consumers to pay higher prices for sustainable choices. In a survey of 30,000 consumers in 60 countries, market researcher Nielsen, for example, found that 55% of consumers said they would pay more for goods and services from socially responsible companies. This number has increased since 2011.[216] As we saw in the figures above, the actual consumption patterns do not match—if they did, market shares for sustainable products would be higher than the average of 1-5%, or the high of 18%, that we have across different consumer sectors.

> **TIP**
>
> An article describing which trade-off considerations consumers need to make to choose responsibly between sustainable cotton and recycled polyester.
>
> ✈ inyurl.com/yyqfkxj7

The inconsistency of consumers' intentions and claims compared to their actual behavior is frequently explained by researchers through basically two phenomena: the attitude and behavior gap and the knowledge and behavior gap.

While attitudes are noble and directed towards sustainability, actual consumption choices are less consistent with attitudes. The sustainable choice at the moment of the purchase decision may not be attractive enough or too costly and inconvenient—or simply out of scope, as it is not customary to consider it.

The knowledge and behavior gap, in contrast, is rooted in too little knowledge, as the name suggests. Consumers do not know the consequences of their behavior or do not know enough about which is the superior option from a sustainability standpoint. Although there is an abundance of information available today, it is still often unclear which is the most sustainable option. Frequently, "it is complicated", as the conscious shopping dilemma example in the reading tip illustrates, to find the right choice, even when it is between sustainable cotton and recycled polyester. It requires some extensive knowledge-gathering, personal preference considerations, an eventual decision, and a change in behavior. That is not an easy journey, and people can understandably get lost easily at times along the way and lose motivation.

A TOUGH ONE: INCONSISTENT BEHAVIOR

Let's deep-dive a bit more into inconsistent behavior, which is a variation of the attitude and behavior gap. To illustrate it, I would like to quote an article from 2008 and its supporting research. Although this article was written several year ago, the phenomenon it describes is still very current in my experience. The research also very vividly describes some inter-relations and typical mindsets.

The study from the British Exeter University interviewed "green idealists" on their attitudes and behavior. This brought biased behaviors to light, causing the researchers to claim that green living is something of a myth. They found, for example, that the longest and the most frequent flights were taken by those who were most aware of environmental issues, including the issue of the threat of climate change. So, the problem, in this case, was not a knowledge gap. The research showed that green lifestyles at home and frequent flying were linked to income. The wealthier people were, the more likely they'd be to engage in both activities. The "green idealists" would basically transfer responsibility to others—for example, by expecting new technology to make aviation greener—in order to not to have to restrict their own comfort or consumption.

The researchers explain: *"People who believe they have the greenest lifestyles can be seen as some of the main culprits behind global warming ... people who regularly recycle rubbish and save energy at home are also the most likely to take frequent long-haul flights abroad. The carbon emissions from such flights can swamp the green savings made at home ... There is this middle-class environmentalism where being green is part of the desired image. But another part of the desired image is to fly off skiing twice a year. And the carbon savings they make by not driving their kids to school will be obliterated by the pollution from their flights. Some people even said they deserved such flights as a reward for their green efforts ... The findings indicate that even those people who appear to be very committed to environmental action find it difficult to transfer these behaviors into more problematic contexts ... The notion that we can treat what we do in the home differently from what we do on holiday denies the existence of clearly related and complex lifestyle choices and practices. Yet even a focus on lifestyle groups who may be most likely to change their views will require both time and political will. The addiction to cheap flights and holidays will be very difficult to break."*[217]

This challenging text indicates that sustainability as a lifestyle topic doesn't keep up. Consistent behavior is hard—to get better in it, we must frequently review our actions and attitudes, but also orientate on truly relevant effects.

ON THE HOOK

Frequently, improvements are out-leveled by more consumption. Such rebound effects (see page 60) cause the increasing emissions in US transportation although cars became more efficient and thus, less environmentally damaging. But as people drive more and buy extra cars, the additional emissions outpace the efficiency gains. Increasing consumption and higher demands can be observed across many countries: In transportation, demands on individual mobility and flexibility have risen. People travel more and longer distances.

But this holds true also for other consumption areas with high impact: There are higher expectations in living space area per capita and interior. In food, trends for convenience, availability, variety, freshness and exclusivity curb resource intensity, emissions and waste. Overall, convenience and higher pace, such as in fast delivery for online ordered items have been increasing for years. The number and variety of consumed items have increased—such in electric appliances and mobile devices, fashion and all kinds of household and spare time appliances. This is about emission intensity of consumption but also about general levels of resource intensity. Even if we use low-energy household appliances, they still contain many resources. Our overall trend to fast consumption and an always-available of goods and services creates a cycle that often leaves out environmental and social impacts.

The general tendency to "more" can be observed at many different levels. We earlier described it as the narrative of our time (see page 32). Consumers are on the hook of this narrative to ever consume more to satisfy their needs. This is a major underlying factor for people to behave unsustainably in day-to-day routines, despite good intentions. According to psychologists and social scientists, individual behaviors are deeply rooted in social and institutional contexts. We are unconsciously influenced by what others around us say, do and which institutional rules are set when we make a choice on our own.

Despite the described positive narrative about sustainability, our reality is marked by endless consumption. It surrounds us daily and basically everywhere. We frequently get mixed messages, more sustainability aspects even in advertising, but still an overall call to really need that item. Bringing up good intentions and living them out in a context that doesn't profoundly nurture them is hard. Additionally, the tension between our reality and the image of sustainability that is drawn creates an imbalance in our daily life. It can be sensed in many facets that an authentic match of rhetoric and behavior, of attitudes and behavior, is lacking.

I believe that the vicious cycle of consumption can only be interrupted when we recognize it and begin to change it step-by-step: more consumption does not deeply satisfy our desires—at least not the underlying longing we are trying to address through consumption, such as recognition, value or security.

WANT TO LIVE ON A SMALLER FOOTPRINT?

How can the personal carbon footprint be reduced? And is it equal to the environmental footprint? Where should the individual consumer focus on?

THE GOAL: 2.3 TONS PER YEAR

"In order to keep global warming at a sustainable level, the international community agreed on a goal to limit the average rise in Earth's temperature to 2°C above the pre-industrial level. In order to achieve this goal, ... each person on Earth has a yearly budget of approx. 2.3 tons of CO_2."[218]

2.3 tons of CO_2 each year—that's the climate compatible annual emission budget for each person on Earth. Most likely, this carbon footprint goal is a personal challenge for you as it is for me. A return flight Frankfurt-Dubai or New York-Los Angeles eats up the yearly carbon budget, one year of average car driving almost does. There are people around the world who live on that budget or even much below, as the first two rows in the table show. And numerous people across the world live far above this yearly budget, as the lower rows show.

Per capita emissions per year, 2018

Burundi	Afghanistan	Haiti	Kenya	Bangladesh	Cambodia
0.027	0.321	0.321	0.374	0.513	0.662

Angola	Congo	India	Bolivia	Colombia	Costa Rica
1.037	1.048	1.833	1.851	1.582	1.659

Brazil	Croatia	Chile	Denmark	China	Germany
2.355	4.169	5.003	5.855	7.717	9.700

Czechia	Japan	Barbados	Iceland	Russia	Estonia
10.336	10.360	11.103	12.228	12.257	13.660

USA	Luxembourg	Australia	Canada	United Arab Emirates	Kuwait
15.741	16.352	16.452	16.855	21.574	23.486

Source: EDGAR 2019.[219] For each country, fossil CO_2 emissions from all anthropogenic activities except land use, land use change, forestry and large-scale biomass burning are displayed.

The average German, for example, would need to reduce emissions by at least 4% per year, or 25% over 10 years, to be in line with the climate compatible emission budget until 2050. An Australian, a US American or a person from Luxembourg would have to double these efforts. When limiting global warming to 1.5 degrees, as most scientists suggest, the global carbon budget would even be only 1.5 tons per year.[218]

Let's be clear: If we believe climate change is a problem, it is also about you and me. There are things politicians and companies and municipalities and other countries have to do. But each of us is part of that journey.

WHAT'S YOUR PERSONAL FOOTPRINT?

A good start is to calculate your personal carbon footprint with one of the good available calculators. Carbon footprints measure how much CO_2 emissions you use by the amount of electricity you use, considering if you use a renewable energy provider, what size, age, style and state of modernization your home has etc. —as all this significantly impacts the personal carbon footprint. Mostly, "CO_2 equivalents" are displayed—this means other greenhouse gases such as methane are converted into CO_2.

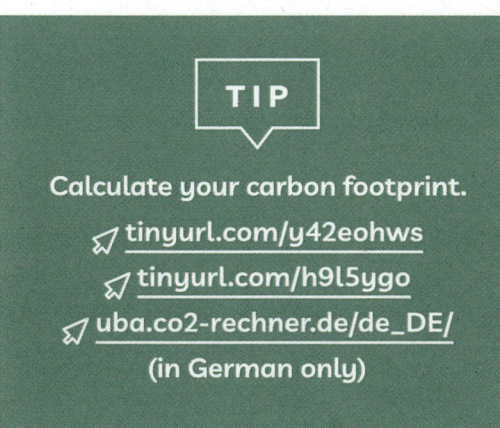

TIP

Calculate your carbon footprint.
tinyurl.com/y42eohws
tinyurl.com/h9l5ygo
uba.co2-rechner.de/de_DE/
(in German only)

Environmental footprints, in comparison, include the so-called biocapacity associated with consumption. That's why some calculators display how many earths we would need if each person lived like you or me. We displayed this perspective when comparing countries (see page 311).

Focus and granularity of the different calculators vary. Methodologies differ, such as in the scope and the emission factors applied. Numbers of average footprints can therefore have a broad range. The mentioned calculators provide good orientation for personal levers which depend on living situations and lifestyle, and thus, they are of highly individual. Especially

in the important area of living, conditions differ widely, according to the housing situation, the way of heating and energy use, the insulation, and so on. For this, the different calculators that include a detailed analysis of the housing conditions are recommended to identify specific levers.

TIP

Calculate your environmental footprint.

footprintcalculator.org

footprint.wwf.org.uk/#/

ecologicalfootprint.com

Over the past years, data availability has increased significantly. For example, when comparing emissions from a flight, calculators such as www.atmosfair.de/en display now the data on average and compared to the most-CO_2-efficient airlines. This transparency is good for consumers but also for corporations in competing for higher efficiency.

HOW TO HANDLE INTERNET USE?

At home, an increasingly important and often underestimated contributor to the carbon budget is the use of the internet in all varieties—mails, social media, searches, streaming and so on. To account emissions to a single flight is today relatively easy. To assign emissions from servers, data warehouses, networks, new fast wire cables and so on to our daily actions is a lot more complicated.

This also holds true because of the enormous pace of new services, offers, data uploads, new providers, which increases fast and opaquely.

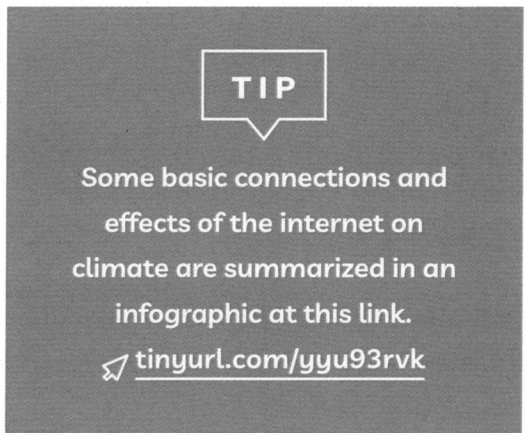

TIP

Some basic connections and effects of the internet on climate are summarized in an infographic at this link.

tinyurl.com/yyu93rvk

Over the past years, across blogs and newspaper reports, the effects of sending an e-mail, making an internet search, streaming half an hour of Netflix or uploading videos to YouTube became increasingly a topic. The efforts to put specific numbers on specific actions or to get a view on the overall effect is noble and necessary—in this dynamic, however easily outdated and in the complexity easily attackable. That´s why I want to focus on a few thoughts across the overall lines of the topic.

In March 2019, a major energy provider was cited in a German newspaper saying that, according to its estimates, the worldwide energy consumption for streaming would be enough to supply energy to all of the households in Germany, Italy and Poland for one year.[220] I cannot easily verify that number, but it sounds big, especially as streaming has risen even further since that time. A 2014 study on streaming emissions from The Shift Project was widely quoted in the press but later quoted of using inflated numbers, in light of other factors such as gains in efficiency that can be achieved through technological development. But what is our conclusion, as consumers, based on these numbers?

TIP

An article for people who want to get into numbers and context of carbon on the internet that gives some factors and forecasts to consider.

tinyurl.com/w3wl9jq

One statistic circulating says that instead of streaming Netflix or Amazon Prime for one to two hours daily, you could let a refrigerator run for half a year.[221] In other reports and analysis, streaming is compared to driving a car, and the numbers range from driving 200 meters or four miles.[222]

Apparently, it is a complicated calculation, depending on multiple variables and assumptions. But what is actually the point? Do we feel better if it is just 200 meters that are emitted? Would it alter anything? Does a single e-mail emit 4 grams but with a huge attachment up to 50 grams? Is it just 0.2 or as high as 7 grams of CO_2 for one Google search? Would it make any difference to you or me if it were double or half that number?

It is necessary to consider the effects the internet has on climate change. Also, we as consumers need to step up to our responsibility in this area, as in all other areas of consumption. Especially, as this is an area that is rapidly growing and our own contribution and tactics for consuming consciously are not yet that clear. Let me therefore provide a few hints on how to manage better in the area of internet use.

 **TOOLBOX**

SIX RECOMMENDATIONS FOR MORE SUSTAINABLE INTERNET USE

– As in all areas of consumption, make sure to frequently consider the following: What is necessary? What can be avoided or reduced? Two easy-to-do tips: Do not send unnecessary messages via e-mail or social messengers, especially not to large groups. Unsubscribe from all newsletters and similar subscriptions that you don't really use.

– Keep in mind that the prevailing view today is that streaming and searching via smartphones, tablets and laptops (in this order) need significantly less energy than via computers and TV. Streaming over 4G mobile networks consumes more electricity than over WiFi.

– As a significant contributor to the overall internet footprint, take into account your devices as well as your respective consumption patterns.

– Make your own contribution by switching to a green energy provider. The tremendous impact of the internet could be partially offset if all providers switched to renewable energy.

– Reduce energy consumption at home, unplug, hibernate, turn off, and charge only if necessary.

– Carbon-offset your internet use, for example, by using internet search engine Ecosia where searches are automatically offset. The following link shows some offset agencies that offer to offset your internet use. ⌁ tinyurl.com/y2ukumpn

YES, DIET NEEDS TO BE MENTIONED

"A Vegetarian's foodprint is about two-thirds of the average American and almost half that of a meat lover. For a Vegan, it is even lower. … eating chicken instead of beef cuts a quarter of emissions in one simple step."[223]

TIP

Just put ⌁ www.ecosia.org into your browser—for every 45 searches, a tree is planted.

Much has been said and written about the impact of different diets and the climate impact of different products.

Plant-based diets versus meat-based diets are the subject of heated arguments among advocates of each group. Let's summarize the basic arguments from a recent scientific study: *"Although the uncertainties involved in data and methodological choices are significant for estimating the carbon and water footprints of agricultural products, there are trends that are suggestive of what products and diets are more environmentally friendly than others, regarding climate change and water scarcity. Results of many studies suggest that carbon footprints of animal products are generally larger than carbon footprints of vegetal products, even without considering the emissions resulting from induced land change. Since the carbon footprint of diets depends on the carbon footprint of individual products, plant-based diets also tend to have*

TIP

A comprehensive statistic about the environmental impacts of food in "Our World in Data".

⌁ tinyurl.com/yadzxa49

smaller carbon footprints than animal-based diets ... Similar results in water footprint estimates suggest that, even without considering many forms of water pollution caused by animal production, animal products are generally more water-intensive than vegetal products. In the same line, studies that estimate the water footprint of dietary scenarios indicate that vegetarian diets have the highest potential to reduce dietary water footprints. The main reason why this pattern is evidenced in the case of carbon and water footprinting, and perhaps on all environmental aspects considered altogether, is just that producing animals for food is inefficient from an environmental perspective. The impacts of animal products and animal diets always include the impacts of the agricultural crops and pastures produced to feed farmed animals, and on top of that the impacts caused by the animals directly, including pollution from manure and enteric fermentation. Therefore, from an environmental perspective, it's always more efficient to eat vegetables directly, instead of giving them to the animals, and then eating the animals ... Animal products and animal-based diets not only release more GHG emissions, and use and pollute more water, but also seem to use and degrade more land than vegetable products and plant-based diets." [224]

TIP

A climate change food calculator compares the environmental footprint of food and drink items based on how often you consume them. It displays comparisons for example to car driving or showers taken.

⟨ tinyurl.com/yyhrwjk5

The factor not included in the study—land use—was analyzed in detail for Germany in another study. The conclusion: Almost half of all land use is needed for livestock diets, while only about 25% of land use goes into crop-based diets. [225] This again favors plant-based food.

The bottom line is that, food is responsible for between 15 to 30% of our personal carbon budget as consumers. Within that budget, livestock adds about 60-80% to our carbon budget, whereby beef and lamb have the highest carbon footprint. Reducing the consumption of livestock products (meat, dairy products, eggs)

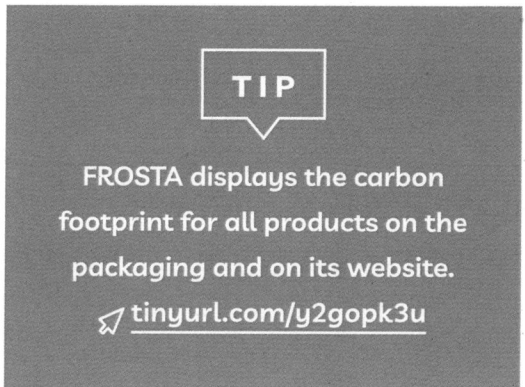

TIP

FROSTA displays the carbon footprint for all products on the packaging and on its website.
tinyurl.com/y2gopk3u

makes a significant difference in terms of climate and overall environmental effects. Reducing meat and increasing the proportion of vegetarian and vegan eating would be by choice. The water and carbon footprint of plant-based products varies widely; avocados, for example, have a much worse balance than tomatoes in terms of emissions per serving and in comparison to fruit, such as berries, grapes and bananas. Cooking with raw, fresh foods instead of manufactured foods is the better choice. Other options to consider include buying regional and seasonal foods, labeled products such as organic and other preferences. I have already mentioned the packaging issue several times, and it is, of course, an issue that is also very relevant in the area of food.

As a general rule, the balance of frozen food is worse than fresh foods because of the high energy intensity needed to prepare and store them. But here, too, it is a question of the details—the balance also depends on the specific type of product and the comparable method of preparing and storing it. FROSTA, one of Europe's leading manufacturers of high-quality frozen food, calculates the carbon footprint for all of its products. The carbon footprint of their frozen products is often superior, especially due to their overall approach of sustainable sourcing, low-carbon ingredients, the freezing method used and energy-efficient cooking in manufacturing.

In food, many people know today the basic facts about impacts. Overall, it is less likely a knowledge but a behavior gap.

There is an additional perspective related to demographic change (see page 209). In a recent study, A.T. Kearney concludes: *"44% of today's global agricultural production ... would be enough to feed most humans. ... a plant-based diet would not only provide the same calories but also have the same nutritional value if crops are chosen accordingly to have enough protein. Hence, we could feed around twice as many humans with today's global harvest if we did not feed livestock but rather consumed the yield ourselves. Based on the current worldwide population of 7.6 billion humans, we would have food for an additional 7 billion people. The number would increase even further if less of the harvest ended up in biofuel and industrial use or if waste could be reduced.*"[193]

10.05 ——————— APPLY THE "GOLDEN RULE"

As a general recommendation or an overall compass for consumption, a "golden rule" for bringing more sustainability into our personal lives comprises five areas:

Avoid—Reduce—Reuse—Recycle—Offset

Frequently applied, it will lead to a habit of making wiser decisions because it requires us to get informed, evaluate and act consciously despite the trade-offs. Let´s look in detail into these five areas.

WHERE CAN I AVOID?

The best consumption is the one we do not make. Sounds radical? In a way, it is. The good news is that given our current consumption patterns and the narrative we are caught up to, it can be a freeing and healthy approach. In our busy, packed lives, reduction, purity and simplicity gain importance.

The potential in our lives to avoid holds true as our lifestyles are fast and full of convenience. Avoiding and reducing often go hand-in-hand. Many things that can´t be avoided can, at least, be reduced. Each reduction or avoidance makes a difference, as they often involve actions we repeatedly do, or they occur in huge impact areas.

Let's illustrate this with an example from the area of mobility that, at first sight, may not look very relevant. *"What if we kept our cars parked for trips of less than one mile?"*

The calculation for the US by the Environmental Protection Agency (EPA) shows that trips of less than one mile add up to 10 billion miles per year—that's the equivalent of the whole population of Chicago driving to Las Vegas and back. Needless to say, that in some cases (like age or mobility issues), avoiding the trip is impossible. Yet, if in the US example just half of all rides of less than one mile could be avoided, the emissions savings would be equal to taking 400,000 cars off the road each year. Plus, millions of dollars of fuel costs could be saved.[226]

TIP

A short pdf from EPA gives details on "What if we kept our cars parked for trips less than one mile?" It also quantifies health cost benefits from avoiding short car rides.

tinyurl.com/yyaqfjxh

In Germany, mobility accounts for about 20% of carbon emissions, with almost 95% created from road travel. About 61% of all trips in Germany are conducted by car. Half of all distances are shorter than 6 kilometers. This means that there is tremendous potential to avoid (or reduce) in this area, too.[227]

As everyone has different lives, routines and needs, and conditions in their respective countries, each person and household have to find their own approach.

Not all types of avoiding always work. There are frequently situations involving a compromise, for example, a last-minute Christmas gift that was ordered online, or using unlabeled fish when labeled fish was unavailable. From my experience, however, it can be helpful to have personal rules or to have made certain decisions in advance that distinguish an area. This makes it easier to install habits for choosing more sustainable options and to change many day-to-day incidences of unconscious behavior.

For example, not owning a car requires me to organize my activities around cycling and public transportation. I also decided many years ago not to fly within Germany, which means—by default—I make my travel plans using trains or buses. Once you have become accustomed to it, expanding the scope of this approach gets easier. For example, I am now al-

AVOID NON-REFILLABLES

Refillable systems are a preferred option for avoiding plastic pollution.

The advocacy organization Oceana published a report in 2020 entitled "Just one word: Refillables. How the soft drink industry can—right now—reduce marine plastic pollution by billions of bottles each year".

A major finding of this report was that a 20% increase in the market share of refillable glass and PET bottles in place of single-use throwaway PET bottles could reduce marine plastic pollution by 39% and keep 8.1 to 13.5 billion PET bottles out of the ocean every year (based on 2018 data).

Of the top 10 global non-alcoholic beverage markets, which includes China, Mexico, Indonesia, India, Brazil, Germany, and Turkey, refillables have market shares ranging between 26% and 61%. Other countries with high market shares of refillable systems include Chile, Peru, the Philippines and South Korea.

Two major markets lacking strong refillable systems, the United States and Japan, sell only 4% of non-alcoholic beverages in refillable bottles. Many countries in the European Union are also at the low end of the refillable spectrum: Finland (2%), France, Sweden, Ireland, and the UK (3%), Denmark, the Netherlands, Greece (4%), Romania, Portugal (5%), Spain (6%). Source and more information: tinyurl.com/yblsurdm

Another approach is to avoid one-way plastic packaging. Single-use packaging is the main method of packaging with a share of over 90% in four regions worldwide (2014 data): Asia-Pacific (38%), North America (21%), Middle East (17%) and Europe (16%).[212] Using refillables saves resources. Liquid soap in a refill bag, for example, reduces packaging by about 70%.

most at the point where I avoid flying anywhere in Europe. My holiday trip to Southern Europe was organized as a progressive tour, where I explored new places along the way. The same holds true for my decision to offset flights. It becomes a lot easier when you set your own clear rules.

Let me share some ideas on how I personally choose avoidance in daily life to illustrate a few points.

TOOLBOX

TWENTY PERSONAL IDEAS FOR AVOIDING

– Reflect before buying any new clothes, shoes, mobile devices, furniture or other bigger purchases: Do I really need to have it? Why?

– Turn off lights, switch off appliances instead of leaving them on standby mode, don't charge devices daily, and avoid "energy vampires" by unplugging chargers and devices.

– Avoid needless heating (for instance, I only heat the living room and the office, and only when I use it).

– Consciously avoid any water waste, switch off taps and use less water for cooking and when showering, etc.

– Avoid conventional energy and switch to a green energy provider that invests in new capacity.

– Avoid flights where possible, limit long-distance flights, and carbon offset any flights taken.

– Avoid owning a car, substitute short-distances car rides with a bicycle or bus ride, or walk.

– Avoid "to go" (coffee, carry out food etc.), especially to save waste.

– Avoid plastic bottles, non-refillables and alumninium cans.

– Avoid wrapping wherever possible (supermarkets, shops).

– Avoid food waste by planning shopping and cooking accordingly.

– Avoid buying the "second" or "third" of something.

– Avoid online shopping.

– Avoid buying items for single occasions (parties, leisure …), instead try to borrow them, trade for them or forget about them altogether—most items aren´t really necessary.

– Avoid owning household appliances—for example, instead of owning a washing machine, share one; avoid items such as dryers, rice or egg cookers. I completely avoid machine-drying.

– Avoid fruits and vegetables out of season and imported items when possible.

– Avoid unlabeled fish, coffee and other food items.

– Avoid softeners and aggressive cleaners, such as drain cleaners.

– Avoid gimmicks, advertising items and any "extra stuff" not needed that usually gets quickly thrown away.

– Avoid small packaging and use refill bags (soap, cleaners) or big packs (coffee), where possible.

WHY REDUCING IS AN OPTION

Most decisions of avoiding have the option of reduction. In our rich countries, we often have to decide things like: How many times a year should we go on vacation? How many weekend trips should we take to see friends or family? How many T-shirts, trousers or dresses should we buy? How often should we replace the mobile device or buy an extra one for convenience? Should we buy a second car or third TV, different bikes for different occasions? The potential for reduction is probably even greater than that of avoiding. When making bigger consumer decisions, such as buying a car, planning a vacation or buying new appliances, it can be helpful to consider the environmental and social effects step-by-step and go for the more sustainable choices.

TIP

"Diet is no longer a private matter" is a provocative statement in the insightful "Meat atlas—facts and figures about the animals we eat" (2014).

foeeurope.org/meat-atlas

It impacted me profoundly when I learned while traveling that many people who work in tourism save their money so that they can go home every three years to see their families. Every three years. I don't want to be misunderstood: I am grateful to live in Germany in a situation that is much more privileged than for many people in the world. I know our lives cannot be compared to the lives of others.

Yet, because I am so privileged, I frequently ask myself: Is there a better way? How can I contribute based on what I have been given? Where are my standards out of touch or harmful to others? I believe there is a responsibility attached to my consumption. It matters how much and how consciously I consume and where I spend my money.

Like avoiding, reduction is a matter of personal preferences, and I believe in freedom of choice. I would like to share a personal example. In reading the list above, I can imagine that some people would have added "Avoid meat and dairy products" due to the significant footprint of these items. In my case, these items are on my reduction list—not because I am unaware of their effects, but because I like dairy products and sometimes enjoy a good steak. Most likely since I lived in Argentina. I have a hard time completely refraining. My approach is to try to reduce my impact in this area, consider the choices and, when I do decide to eat meat or dairy product, really enjoy them.

It's often about daily routines and small things. For example, I frequently observe people unconsciously using four to five paper towels to dry their hands in public restrooms or leaving the water tap running. Imagine if each of these people would use just one or two paper towels less or impress upon their children the importance of conserving water using the example of their own behavior. Daily routines are a great area to explore what can be reduced. Were you aware of the following effects of daily routines?

 **TOOLBOX**

CONSIDER THE EFFECTS
OF DAILY ROUTINES

– Washing clothes at 30 degrees uses around 40% less electricity over a year than washing at higher temperatures, according to the Energy Saving Trust.

– A 5-minute shower is the recommended length of time if you shower daily. Each one-minute reduction in your daily shower can save between 547 and 2007 gallons (2,070–7,597 liter) of water per year.[228]

– LED light bulbs reduce energy consumption by over 80% and last up to 25% longer compared to conventional light bulbs. If every person in the US replaced one conventional light bulb with an LED, enough energy would be saved to light 2.5 million homes or offset the greenhouse gas emissions from 800,000 cars.[229]

– Reducing meat consumption is one of the changes that has the big impact. A meat-free day reduces impact by 15%. Choosing chicken instead of beef drops the footprint fivefold. Reducing the size of a rib-eye steak from 6 oz to 4 oz reduces the carbon footprint by half.[230]

– Approximately 20% of the total electricity use in buildings world-wide today goes towards air conditioners and electric fans. The International Energy Agency expects space cooling energy needs to triple by 2050.[232] According to a recent study, emission reduction from different and more efficient air conditioning is tremendous.[232]
Any reduction in the use of an air conditioner or replacing it with more efficient one makes a difference.

Finally, with regard to reduction, I would like to raise a personal point as it is a part of our culture and especially common in circles with good incomes to spend. Bringing along a gift to every dinner invitation, coffee among friends and drop ins of all kinds. Giving kids a material reward for each small achievement. Buying stuff because it's Christmas, Easter, Valentine's day, a birthday, bachelor(ette) party, baby shower, an anniversary or for pre- and pro celebrations of all kinds. Don´t get me wrong, I love parties and community, and I like making gifts. Increasingly, however, I see it has become an inflationary routine and a social obligation. This can lead to a vicious cycle of ever more pressure for the giver and little appreciation and value for the receiver—on top of the negative effects all of the "stuff" we buy to give away has on people and nature.

In the area of gifts, I intentionally and deliberately reduce: I set myself a rule to consider gifts carefully instead of bringing along something by default or making a gift for all of the occasions where it is socially expected. People who know me well know that I love to give. Thus, when I do, they know it's from the heart. I try to reduce buying "stuff" that's not needed and opt for alternatives instead whenever possible. For example, home-made jam or pesto, a hand-written card or some seasonal flowers, a gift of family time together. When it comes to children, I also try to give time instead of things. For instance, reading them a story, doing arts and craft, going to the playground or cooking with them. It's a comparably small area of my footprint. It is not easy to implement as it is against my natural intentions (of giving), and it likely disappoints expectations. I reduce because I think it is a good step in curbing and challenging our narrative that more consumption gives us what we truly want.

I believe that avoiding and reducing is about making wiser decisions and intentionally changing habits to reduce the negative impact on the environment and people. It is a lifelong process that each of us can initiate freely and without pressure, originating from a sense of responsibility and a heart that searches for positive change.

THE FUN WORLD OF REUSING

Many things in our convenient lives are made for a single use. That's why avoiding and reducing are so important. To reuse means to give a longer lifespan to products once they have been bought and before they go into the garbage for recycling or combustion. It takes some consideration to come up with where and how to reuse.

A simple example is the reuse of plastic or paper bags for groceries from supermarkets. Many people take these bags home and throw them away immediately, as we frequently do with other packaging. While the best solution would be to avoid using these bags altogether, reusing a bag at least one more time multiplies its value. Bags can be reused for packing or carrying things or collecting garbage instead of buying bin liners.

A typical example is the common sharing of baby and children's clothes. In this case, an economic benefit facilitates the reuse: As clothes are relatively expensive and have short lifespans because babies and small children grow so fast, they are frequently re-distributed among friends and neighbors. This means they are used up to four or five times before their end-of-life. The self-organization and high level of reuse in this area could be used as a model for other areas of consumption.

The use of the internet has made reuse a major market, especially among private people. Platforms make it easy to match supply with demand for all types of items that people collect or use in their daily lives, from sofas and tables to curtains and washing machines. New applications, such as platforms specialized in books, music, antiques or other kinds of products, have multiplied.

Reuse in the form of sharing has become a major trend, facilitated by the internet, too. Sharing in the area of mobility services and private rooms are increasingly used. Not yet common is the sharing of household appliances, such as washing machines, garden equipment or tools. Reuse through sharing in this area has tremendous potential, for example, sharing among neighbors.

Several items can be very valuable to other people. For example, poor people living in cities in rich countries could benefit from warm clothes, a new pair of shoes, toys or school bags for their children donated by people in their immediate surrounds. Establishing connections to charities that work across the city with people in need gives a good feeling to know that valuable use is being made of items while reducing waste in landfills. The reuse of food reduces the waste of valuable resources and matches them with people in need. Food banks across the world or apps increasingly fill the gap.

TIP

An app to share food and other household items for free.

olioex.com

Reusing should be carefully considered for all types of electronic devices and electric appliances. There certainly is a huge demand in less developed countries for these products, and we would probably feel good about giving away our old printers, screens, TVs and refrigerators, smartphones and laptops for free—so someone else could reuse them. Private collection initiatives sometimes call for exactly that. As there is currently a major problem with the end-of-life for most of those products (see pages 284ff.), it should be ensured that devices are disposed of properly. This could be a better alternative to properly recycling items at local recycling stations.

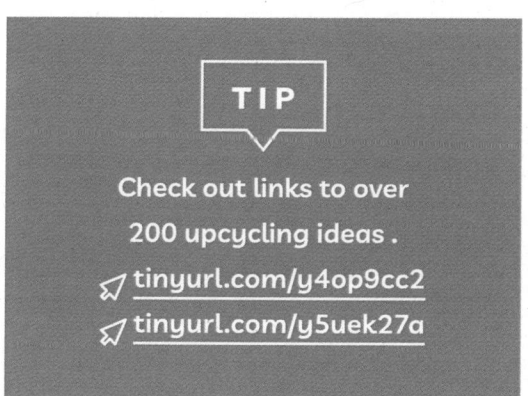

TIP

Check out links to over 200 upcycling ideas .

tinyurl.com/y4op9cc2
tinyurl.com/y5uek27a

The trend towards reusing finds creative expression in upcycling. It is the process of transforming products or waste materials into new products that are often perceived as valuable through their creative value. Upcycling workshops in metropolitan areas often get a tremendous response from young and creative people.

RECYCLING MATTERS, PERIOD.

A key element of recycling is proper disposal, which allows the process of converting materials for reuse to avoid discarding them. Let's look at some data.

According to the 2018 figures of the United Nations, 11.2 billion tons of solid waste is produced on Earth every year, contributing to about 5% of all global greenhouse gas emissions, as quoted in a post. The top 5 recycling countries were listed as Germany (56.1% of the waste produced is recycled), Austria (53.8%), South Korea (53.7%), Wales (52.2%) and Switzerland (49.7%).[233] Another report mentioned an overall higher recycling quota with Austria ranking first (63%), Germany second and Taiwan and Singapore in the top ten.[234] Many developing nations lack any good recycling systems.

Systems across the world differ generally and are difficult to comparable. However, successful countries usually have some regulatory practice implemented as well as logistics that support individual recycling efforts. Germany has a collection system for glass (white, brown, green) with over 250,000 containers across the country that has been in place since 1974. This facilitates the process for consumers and the glass recycling industry. Germany recycles 85% of its glass [235] compared to the US which recycles just one-third.[236] Glass is among the materials that can be 100% recycled and reused repeatedly, if correctly sorted and collected.

TIP

An informative video on glass recycling.
tinyurl.com/y35b8scl

Closed loops—recycling over and over again, also holds true technically for many materials manufactured in electronics, if properly collected and recycled. In the US, more than 130,000 computers are thrown away each day and cell phones contribute 65,000 tons of electronic landfill waste every year. Yet almost the entire computer can be recycled: plastic, metal and glass; and phones have valuable materials that could be recycled.[237] Metals (steel and aluminum) have a relatively high overall recycling quota. Steel is said to be the world's most recycled material, with a recycling rate of 86%

worldwide (2014 data[238])—this is due to many industrial processes and mandatory recycling quota for example for automobiles. Aluminum is easy and cost-effective to recycle. China is the world leader in reusing aluminum, with 99.5% of its aluminum waste being recycled; worldwide it is at about 69%.[239] With respect to aluminum cans, Brazil (96.9%) is the country with the highest rate of recycling, followed by Japan (93.6%), the EU (74.5%) and the US (63.6%).[240]

Great logistic efforts are involved and often huge amounts of energy, water, and some emissions are associated with recycling processes. However, for many materials, it is still cost-effective and environmentally better to recycle than use virgin materials. Recycling paper, for example, uses significantly less energy (roughly 28 to 70% less) in comparison to producing paper from virgin material. Paper fiber can be recycled and reused up to 7 times.[241] The energy saved from recycling a single aluminum can, or one glass bottle, could be used to power a television set for three hours, or light a 100-watt light bulb for four hours.[229]

TIP

An informative article on paper recycling, explaining among other things environmental effects and which sorts of paper are usually not recycled.

tinyurl.com/y5ucmfem

For the textile industry, there is still no global solution. Only 1% of textiles are recycled into the same or similar quality applications. 12% are recycled into lower-quality applications and 73% are landfilled or incinerated.[242] This situation is also a result of the increasingly lower quality of textiles. In the light of the increasing amounts of textiles produced and consumed (see pages 274ff.), this is another part of the sad apparel story which will hopefully be changing soon.

The problems associated with plastic were already mentioned and also manifest at the end-of-life. Worldwide, 400 million tons of plastic are produced each year. Packaging—usually single-use—accounts for more than one-third of the plastic produced.[212] Since the beginning of mass-producing synthetic materials in the 1950s, 8.3 billion tons of plastic have been produced worldwide, of which only 9% were recycled. Still today, only 14% of plastic is recycled, mostly downcycled—making products of less value.[212] According to this analysis, Germany, otherwise among the recycling world champions, recycles only 15.6% of plastic. Additionally, plastic waste doubled between 1995 and 2017. The current per capita use of plastic packaging is at a high 38 kilogram per year, with higher per capita use in the European Union only in Luxembourg, Ireland and Estonia.[212]

TIP

A recently published "Plastic Atlas" (available in German only) gives a profound overview of the problem and the approaches to solving it. Refraining from and reducing the use of plastic are the single most powerful actions consumers can take.

tinyurl.com/y7zsxh2o

It is estimated that the burden from microplastic in soils is 4-23 times higher than in the oceans. There is still little research on this topic. Residues of microplastic are contained in effluent sludge that is used as fertilizer in agriculture, with the UK, France, Germany and Spain as the top countries in Europe applying it.[212]

According to EPA statistics, in the US, only about 7% of all plastic entering the municipal solid waste stream is recycled.[243] The primary plastic exporters worldwide are the US, Japan and Europe. In 2016, China imported 600,000 tons of plastic waste per month until it posed a ban on this in 2018. Now, Malaysia is one of the countries that receives huge amounts of plastic waste.

As recycling systems differ across the world, universal recommendations are not easy to make. Each person needs to apply recycling to fit their own individual context and lifestyle. Yet, some of the following recommendations might hopefully still give some guidance on the issue of recycling.

 **TOOLBOX**

CONSIDERATIONS FOR YOUR OWN PERSONAL RECYCLING

– Recycling starts with consumer decisions. Refrain as much as possible from using materials that do not perform well in the recycling cycles of your country.

– Target to reduce the overall amount of garbage.

– Sort garbage.

– Avoid and reduce plastic. Recycle plastic.

– Recycle glass.

– Separate lids from glasses and other packaging.

– Recycle paper and use recycled paper products.

– Recycle all types of waste containing hazardous substances such as light bulbs, batteries, paints and household demolition waste (including carpets, building materials and so on) in specified collection centers.

– Recycle electric appliances and electronic devices in specified collection centers. Some mobile device manufacturers have installed collection and recycling programs you can use.

– Get informed about recycling organic waste such as peels or leftovers. In some cases, it is better to recycle this waste in the garbage or in organic waste containers.

– Reduce textile waste through less consumption. Find out about adequate recycling of textiles in your area.

FEELGOOD OFFSET

Compensating emissions or carbon offset, as it is also called, is an easy way to compensate for the emissions that result from daily life or for especially carbon-intensive activities, such as flying and social events. How does it work? Atmosfair, a German organization specializing in carbon offsetting, describes it using flight compensation as an example:

"Airline passengers make a voluntary climate protection payment based on the amount of emissions they create; atmosfair uses these contributions to develop renewable energies in countries where they hardly exist, above all in developing countries. In this way, atmosfair saves CO_2 that would otherwise be created by fossil fuels in these countries. Meanwhile, local people profit since often for the first time; they gain access to clean energy available around the clock, which is a must for education and creating equal opportunities."[244]

As the name suggests, carbon offset compensates your CO_2 emissions but no other environmental or social consequences of consumption. Yet, with climate change being a major sustainability issue, it is a highly relevant step. It is easy—just costs a few clicks and some money. For example:

Activity	CO_2	Offset
Flight from Hamburg to New York, roundtrip, economy class	2,875 kg	67€
Flight from Hamburg to New York, roundtrip, business class	5,391 kg	124€
Flight from Hamburg to London, roundtrip, economy class	349 kg	10€
Car ride 12,000 km per year, middle-sized vehicle	2,000 kg	46€
Average yearly footprint of a German consumer	11,000 kg	253€

Source and origin for further calculations: Atmosfair website. www.atmosfair.de/en/offset/flight/. Accessed November 28, 2020.

A key factor in offsetting well is to partner with trustworthy organizations. Look for the following criteria when searching for such organizations:

– The organizations and/or the projects are certified.

– Compensation projects are additional, new projects.

– There is a high level of transparency, for example, with respect to calculation methods as well as to which projects benefit from the payments.

– The majority of the payment goes into the project itself.

It is useful to check the organizations and their compensation approaches as they differ widely in quality. Often, certificates are bought only to pay off the emissions created but not to invest in making real change. This is one reason why climate compensation does not have a good reputation. Frequently, it is used in combination with the word "Ablasshandel" (selling of indulgences). This was a practice applied in the medieval age by the Catholic church by which people paid money to receive forgiveness of sins from the church. The system came to an end with the reformation of Martin Luther and the "sola gracia" and "sola fide"—justification is by grace and faith alone.

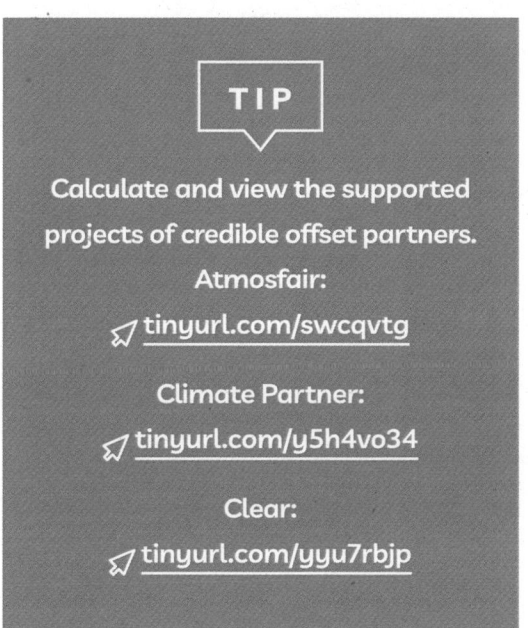

TIP

Calculate and view the supported projects of credible offset partners.
Atmosfair:
tinyurl.com/swcqvtg

Climate Partner:
tinyurl.com/y5h4vo34

Clear:
tinyurl.com/yyu7rbjp

Besides the assumed monetary whitewashing, there is a notion of climate compensation that understands it as a free riding pass. It surely should not be, but instead combined with avoidance and reduction, as atmosfair also emphasizes. They also argue that mere compensation would not make it possible to achieve the limitation of global warming under the Paris Agreement, even if we offset all emissions of rich countries. Yet, it is relevant and, certainly for consumers, a simple and impactful contribution to financially support climate-friendly technologies and economic progress in developing countries.

APPLICATION

1. ——— Take an inventory!
Calculate your eco and carbon footprint with one or several of the calculators. Check the associated tips for making improvements.

✎ ecologicalfootprint.com   A rough estimate of your eco-footprint

✎ footprintcalculator.org   A detailed analysis of your eco-footprint

✎ footprint.wwf.org.uk/#   A detailed analysis of your eco-footprint

✎ uba.co2-rechner.de/de_DE in German   A detailed carbon inventory, with scenarios for improving short- and mid-term and until 2050

✎ tinyurl.com/y42eohws   A detailed carbon inventory

✎ tinyurl.com/h9l5ygo   A detailed carbon inventory, especially at home, US-focused

2. ——— What are your personal key takeaway's from the chapter on the role of consumers and impact levers?

3. ──── Guess which countries have which carbon footprint. Match numbers with countries. How big was the annual climate compatible émission budget for a person a) in the 2° degree scenario? b) in the 1.5° degree scenario?
Emissions per capita per year, in tons of CO_2

Albania	1.037
Algeria	1.649
Angola	1.715
Antigua and Barbuda	1.852
Argentina	3.321
Armenia	3.871
Aruba	4.743
Australia	6.118
Austria	8.271
Azerbaijan	9.115
Anguilla	16.452

Data for 2018. Source: Edgar 2019 [219]

4. ──── Which three things would you like to start changing?

5. ──── What questions are still open for you at this point? Which topics do you want to deep-dive more into? How could you approach it?

HOW TO GO ON

We've taken a bit of a journey together—thanks for being onboard with me. We navigated sustainability across its overall understanding and the different industries, and deep-dove into some specific issues and contexts. You were invited to reflect on applying these insights to your own context and hopefully learned a few new things along the way.

Now, the journey is over. Or is it? No, it is definitely not. While this book may be over, the journey starts right here. For many of you, it will continue and do so for a while. I can promise you that! There is so much to do, to change, to invent, and to shape before we finally come to the point where we will no longer need special books about sustainability. When we know what it is about and we navigate it as a natural part of life—like walking and talking. Hopefully, on your journey, the book's insights will give you new inspiration now and again.

I thought about finishing this book with a summary, an outlook, some smart thoughts on future perspectives. I refrained from it. It's all there already. If we consider the food for thought in this book and if we all implement only half of what I have recommended or inspired you to do by 2030, our world would look better and be more sustainable by the end of this decade. And we would be more motivated to shape it further into exactly that direction, as we experience how inspiring this path is.

That's why I would like to finish very simply—with encouragement and a challenge.

The encouragement: Each of us is different as a result of our personalities, background, the places we live, the jobs we do, and the skills and opportunities we have. Each of us can contribute. You make a difference, not only when it comes to the small things in daily life but also in the bigger context. Be encouraged. You have an impact on sustainability. And only you can impact what you have been entrusted to. Your part matters, to you and others.

The challenge: I repeat the wish I told you in my invitation.

Take at least three specific points out of this book— things that touched you, things that give you a sense of urgency or the nagging feeling that this specific point could or should be changed. And issues where you feel: "This is something I am drawn to because it's affecting my particular circle of impact."

You could, of course, think small and make these three specific points something like: 1) I will recycle, 2) I will have a meat-free day a week and 3) I will cut my car rides below one mile in half. I invite you to expand your perspective: Let it not be just about you and only some minor personal limitations—let it stretch you.

Think big, at least in one point out of the three. Challenge yourself and your surroundings. Take, for example, the 2.3 tons per year. Question the status quo. Could there be an alternative to the current growth narrative at your company, for instance? Aspire to influence what you currently believe is impossible. I talked to a friend recently—a doctor who tries to live sustainably at home but is frustrated every day when he sees the tremendous amount of waste and single-use items in the hospital every day. I encouraged him to take it on as a task in his circle of impact where change right now seems just too big and out of reach. Remember: impossible is an attitude. I believe that great things can emerge out of our willingness and readiness to change for the better. It costs us something, but it's worth it. Be blessed in your endeavors.

EPILOGUE

Let me share, at the very end, a personal transformation I went through. It is a tiny insight into my personal journey and my heart—confessing and emphasizing the point that transformation always has to do with oneself, too.

I´ve been involved in sustainability for more than 25 years. This includes the time I worked in the field as well as my studies on the subject at university. One of my core motivations has always been and still is my love for the world and its people. I have traveled extensively and feel deeply touched by the world's beauty. Having explored the breathtaking underwater world as a diver makes me want to fight for its conservation. As part of my personality, seeing injustice and people who are not able to live at their full potential breaks my heart.

A second strong motivation emerged when I experienced that I can have an impact with my knowledge and capabilities to lead others towards change. In numerous sustainability client projects, as well as in my work with students, MBA peers, team members and in professional networks, I felt and experienced that I can have an influence on people and help companies advance in their transformation.

This is rewarding.

Yet, over the years, while working in sustainability and arguing for change and transformation as a professional, I would see three inclinations pop up in myself in my passion for sustainability.

I thought it depended on me to save the world. The impactful work I conducted granted meaning and recognition. It would repeatedly lead to overconfidence and ignited a hunger to get even better, bigger, and more prestigious projects and assignments to literally save the world. Glory, not change, had at times become the main motivator.

I felt superior. I had a nagging feeling of being a bit better than others. I'm on the good side, engaged for the benefit of the world and its people, dealing with topics that really matter, taking action and getting it done. When others don't. Sometimes this may have even been true. But it often made me arrogant, pointing fingers at others, whether openly or in my mind and heart.

I felt frustrated. The complexity of sustainability can sometimes be overwhelming. Battling for change can feel useless, tilting at windmills. The bad world, the egotistic people, double standards, the never-ending inertia in companies got on my nerves. Seeing my own ambivalence in sustainable living in addition made room for feelings of guilt and shame.

While I still have these tendencies repeatedly popping up, I can now recognize them well and have found a remedy for them. It began through a series of coincidences when I began my journey as a Christian believer at the end of 2008. There were a couple of times when I spoke the words: *"God if you exist, reveal yourself to me."* A lot happened, different than I expected, but truly life-changing. Over the next few years, I decided to dedicate myself fully to Jesus. My faith is the foundation of my life and my navigator through all of it. The very wise Augustine (354-430) said: *"Our hearts don't find rest unless they rest in you."* I confess that this is true in my own personal life.

How this has translated into a transformation of heart and how it ultimately links to sustainability is something I probably need to explain.

I discovered through the experiences that revealed the three inclinations I mentioned—some of them very deficient—that I'm so much less noble and consequent when it comes to good intentions and actions than I should and would like to be. I would really love to be a role model and a sustainability saint. I am not. And I would love to save the world with my efforts. I can't.

But thankfully, I don't have to. I have recognized and can now believe from the heart that it's not up to me and to my righteousness but to the life and sacrifice of Jesus Christ that makes the difference. How so?

As my value and security don't depend anymore on prestigious projects, people liking me, or generally on results, I am free and fearless to share who I am and what I know wherever I'm at. For the greater impact generated, God is responsible. And as I don't have to be a 100% perfectly sustainability role model by God's measures, I am allowed to fail. I can newly decide each day and on each occasion out of my freedom what is better for people and nature. This is freeing, empowering and effective at the same time. Additionally, knowing that this world has had an ultimate redeemer already, empowers me to give what I have to contribute to the renewal of this world while I live here, step-by-step in the areas I am in. I found sustainability reveals attitudes and motives and tests character. And I'm ultimately happy that I don't have to live by my own standards or those of other people anymore.

How could this possibly relate to you, at a personal level?

First, I would like to direct a call for action towards my fellow Christian sisters and brothers: We indeed should be at the forefront of sustainability. We have not stepped up and embraced this enough in our work, homes, churches or in our societies. Practicing sustainability is not just a command to steward the world wisely. It also reflects how much the Lord we believe in loves and values this world and its people. I pray that change happens from the inside out, by grace.

To everybody: When I said twelve years ago: *"God if you exist, reveal yourself to me."*, I invited him into my life. Today I know, I was invited first and just responded to it. I want to share and encourage that this is the most valuable invitation, and it is accessible to each person who chooses to be open for it.

HOW TO GET IN TOUCH

I welcome you to get in touch – I would love to know how the book impacted you, which questions arose, where you are on your journey, and how I can likely navigate you through.

Reach out via email for any feedback, questions or requests. hello@sustainavigator.com

Visit me at my webpage for frequent blog posts, free resource downloads, and an overview of my offers:

✈ www.sustainavigator.com

At the website, you can also opt in to receive a quarterly newsletter for inspiration and exclusive insights on corporate sustainability.

Check out social media for updates, promotions and personal insights.

📷 sustainavigator

f sustainavigator

in sustainavigator

✗ Anke_Steinbach

Get support on your sustainability journey–online or offline.

From one-time expertise to long-term accompaniment–it's tailor-made and solution-oriented. You can trust to focus on the right things and achieve assured quality results. Serving my clients includes challenging them so they can confidently expand in their sustainability performance and leadership.

THE MOST REQUESTED PROJECTS INCLUDE

Sustainability review
Know where you stand and what needs to be done

Strategy building
A mid-term plan for your sustainability journey

Communications
Publish a sustainability report or comparable material

SPECIAL PACKAGES INCLUDE

Checkups
(A) Sustainability strategy (B) Website communication
(C) Sustainability Report (D) Benchmarking

1 hour online consulting
single time, monthly, quarterly

Sneak consulting
Get an answer to your specific sustainability question

Sustainability onboarding
Coaching sustainability managers into their task

To find out more about any of these options, contact me per email.

hello@sustainavigator.com

 Sustainavigator

REFERENCES

1 —— World Health Organization (WHO) website. https://covid19.who.int/. Accessed November 28, 2020.

2 —— Interview with Stewart Brand, ZEIT paper issue October 8, 2020, pages 38-39.

3 —— Merriam Webster website. https://tinyurl.com/y2la5sa5. Accessed December 7, 2020.

4 —— Fortune Global 500 website. https://fortune.com/global500/search/. Accessed August 1, 2020.

5 —— Palmer A. Amazon to hire 75,000 more workers as demand rises due to coronavirus. CNBC website. Published April 13, 2020. Accessed December 7, 2020.

6 —— Tata group website. https://www.tata.com/business/overview. Accessed August 1, 2020.

7 —— Acting responsibly. Schwarz group. https://tinyurl.com/yxc82csa. Accessed December 7, 2020.

8 —— Bosch group website. https://www.bosch.com/company/our-figures/. Accessed August 1, 2020.

9 —— US Bureau of Labor Statistics website. https://tinyurl.com/yyf2usko, Accessed December 7, 2020.

10 —— Statista website. https://tinyurl.com/y49d7yh5. Accessed December 7, 2020. 2018 data.

11 —— Statista website. https://tinyurl.com/yymd6dcl. Accessed December 7, 2020.

12 —— How do the world's biggest companies compare to the biggest economies? World Economic Forum (WEF) website. Published October 19, 2016. https://tinyurl.com/yxmqcp9a. Accessed December 7, 2020.

13 —— Most valuable companies in the world–2020. FXXSI website. Published November 2, 2020, data from September 6, 2020: https://tinyurl.com/y3gtfvc6. Accessed December 7, 2020.

14 —— Global 2000. Forbes. Published May 13, 2020. https://tinyurl.com/y6gjke85. Accessed December 7, 2020.

15 —— Unilever website. https://tinyurl.com/y6pzhzr9. Accessed December 11, 2020.

16 —— Amazon website. https://tinyurl.com/y7manddj. Accessed December 11, 2020.

17 —— Volkswagen website. https://tinyurl.com/ycop25mp. Accessed December 11, 2020.

18 —— Novartis in Society. ESG Report 2019. https://tinyurl.com/yys3avfx. Accessed December 7, 2020.

19 —— BASF website. https://tinyurl.com/yy2o3x4m. Accessed December 11, 2020.

20 —— Nestlé website. https://www.nestle.com/aboutus/suppliers.
Accessed December 11, 2020.

21 —— McDonalds website. https://tinyurl.com/y2xopy4x. Accessed December 11, 2020.

22 —— HSBC website. https://www.hsbc.com/who-we-are. Accessed December 11, 2020.

23 —— The Coca-Cola Company 2018 Business & Sustainability Report.
https://tinyurl.com/rv2qq7n. Accessed December 11, 2020.

24 —— The Decade to Deliver. A Call to Business Action. The United Nations Global
Compact—Accenture Strategy CEO Study on Sustainability 2019.
https://www.unglobalcompact.org/library/5715. Accessed December 11, 2020.

25 —— New report shows just 100 companies are source of over 70% of
emissions. Carbon Disclosure Project (CDP) website. Published July 10, 2017.
https://tinyurl.com/y88jplex. Accessed December 11, 2020.

26 —— The impact of multinationals in developing countries. World Benchmarking
Alliance (WBA) report. Published May 7, 2020. https://tinyurl.com/yxmc7vb5.
Accessed December 11, 2020.

27 —— DHL Global Connectedness Index Updates 2018 and 2019. DHL website.
https://tinyurl.com/y38oqrt3 and https://tinyurl.com/y37crdce.
Accessed December 11, 2020.

28 —— Meyer N. I., Nørgaard J.S. Policy Means for Sustainable Energy Scenarios.
Published in: International Conference on Energy, Environment and Health. 2011.
https://tinyurl.com/y6tcp6vh. Accessed December 11, 2020.

29 —— Summary for Policymakers of the IPBES Global Assessment Report on
Biodiversity and Ecosystem Services. Intergovernmental Science-Policy Platform
on Biodiversity and Ecosystem Services (IPBES). 2019.
https://ipbes.net/global-assessment. Accessed December 11, 2020.

30 —— Badger S. The Old Model of Capitalism Needs to Evolve—It's Time for a New
Roadmap. https://tinyurl.com/y3ha5fw6. Accessed November 28, 2020.

31 —— Gilbert J.C. The Most Important List Of Businesses In The World. Forbes.
Published September 12, 2017. https://tinyurl.com/y5rkwqhp. Accessed December 11, 2020.

32 —— Ripple W.J. et al. World Scientists' Warning of a Climate Emergency.
BioScience. January 2020 Vol. 70 No. 1. https://tinyurl.com/y5cm3l4l.
Accessed December 11, 2020.

33 —— From Smart to Senseless: The Global Impact of 10 Years of Smartphones.
Greenpeace report. 2017. http://tinyurl.com/y9c6yrdr. Accessed December 11, 2020.

34 —— Schmithüsen F. Three hundred years of applied sustainability in forestry. Unasylva 240 Vol. 64 2013/1. http://www.fao.org/3/i3364e/i3364e01.pdf. Accessed December 11, 2020.

35 —— Turner G., Alexander C. Limits to Growth was right. New research shows we're nearing collapse. The Guardian. Published September 2, 2014. https://tinyurl.com/znwrtuk. Accessed December 11, 2020.

36 —— UN. Secretary-General. Report of the World Commission on Environment and Development. 1987. https://digitallibrary.un.org/record/139811. Accessed August 15, 2020.

37 —— Crippa M. et al. Fossil CO2 emissions of all world countries. 2020 report. Ispra: JRC SCIENCE FOR POLICY REPORT. 2020. https://tinyurl.com/y5eklj27. Accessed December 11, 2020.

38 —— Ritchie H., Roser M. CO_2 and Greenhouse Gas Emissions. Our World in Data website. First published May 2017; last revised August 2020. https://tinyurl.com/yy29hp2w. Accessed October 28, 2020.

39 —— UN website. https://www.un.org/sustainabledevelopment/development-agenda/. Accessed November 13, 2020.

40 —— Millennium Development Goals. UNDP website. https://tinyurl.com/tvxgufl. Accessed November 13, 2020.

41 —— UN website. https://sdgs.un.org/goals. Accessed November 13, 2020.

42 —— Smith R. E. Defining Corporate Social Responsibility: A Systems Approach For Socially Responsible Capitalism. University of Pennsylvania. 2011. https://tinyurl.com/y5a8rgqs. Accessed December 11, 2020.

43 —— The Walt Disney Company website. https://tinyurl.com/y57dx4vs. Accessed December 11, 2020.

44 —— Water Footprint of products. Water Footprint Network website. https://tinyurl.com/y378ytau. Accessed November 13, 2020.

45 —— Product gallery. Water Footprint Network website. https://tinyurl.com/y3mmww5l. Accessed November 13, 2020.

46 —— Ritchie H. Which form of transport has the smallest carbon footprint? Our World in Data website. Published October 13, 2020, https://ourworldindata.org/travel-carbon-footprint. Accessed December 11, 2020.

47 —— Statista website. https://tinyurl.com/y9tqk72e. Accessed December 11, 2020.

48 —— Der Rebound-Effekt: Störendes Phänomen bei der Steigerung der Energieeffizienz. Deutscher Bundestag Wissenschaftliche Dienste. 2014 Nr. 16/14. https://tinyurl.com/y2powka7. Accessed December 11, 2020.

49 —— Rowley S. Could the rebound effect undermine climate efforts? The Guardian. Published February 22, 2011. https://tinyurl.com/y83xh6pg. Accessed December 11, 2020.

50 —— Microsoft website. https://tinyurl.com/y2u85mun. Accessed December 11, 2020.

51 —— Quote by Alberto Carrillo, Director Science Based Targets at Carbon Disclosure Project (CDP). Quoted in the LafargeHolcim press release on the climate pledge. https://tinyurl.com/y545ca5v. Accessed December 11, 2020.

52 —— Armacell Annual Report 2019. https://tinyurl.com/y6jpgv6h. Accessed December 11, 2020.

53 —— Deutsche Umwelthilfe website. https://www.duh.de/becherheld-problem/. Accessed December 11, 2020.

54 —— Gabbatiss J. Disposable coffee cups: How big a problem are they for the environment? Independent. Published January 5, 2018. https://tinyurl.com/ycyre66h. Accessed December 11, 2020.

55 —— National water footprint explorer website. https://tinyurl.com/y69zfjg5. Accessed December 11, 2020.

56 —— Nestlé Stakeholder Engagement. Nestlé website. https://tinyurl.com/yy994tv5. Accessed December 11, 2020.

57 —— Microsoft Stakeholder Engagement in the Governance of Corporate Social Responsibility. Microsoft website. https://aka.ms/Stakeholderengagement2019. Accessed December 11, 2020.

58 —— Daimler Annual Report 2019. https://tinyurl.com/y68j6pxo. Accessed December 11, 2020.

59 —— Bayer Stakeholder Dialogue. Bayer website. https://tinyurl.com/y2qmhbcu. Accessed December 11, 2020.

60 —— Doyle J. "The Brent Spar Fight, Greenpeace: 1995." PopHistoryDig.com. Published April 15, 2020. https://tinyurl.com/y6kwjmsp. Accessed November 13, 2020.

61 —— Bangladesh raises minimum wages–is it enough? Fair Wear Foundation website. Published September 28, 2018. https://tinyurl.com/y6ldrop4. Accessed December 11, 2020.

62 —— Plambeck E.L., Denend L. The Greening of Wal-Mart. Stanford Social Innovation Review. 2008. https://tinyurl.com/yyz4sna4. Accessed December 11, 2020.

63 —— Global Energy Perspective 2019. McKinsey & Company website. https://tinyurl.com/y6ock3qm. Accessed December 11, 2020.

64 —— Share of renewables in electricity production. Global Energy Statistical Yearbook 2020. Enerdata website. https://tinyurl.com/yaf7em5a. Accessed December 11, 2020.

65 —— PS Market Research website. https://tinyurl.com/y5qysrdk. Accessed December 11, 2020.

66 —— Solar plant the size of San Francisco powers Morocco's sunlit ambitions. Climate Investment Funds. Climate Home News website. Published January 22, 2019. https://tinyurl.com/y4hm4gbe. Accessed December 11, 2020.

67 —— Dudley D. China Is Set To Become The World's Renewable Energy Superpower, According To New Report. Forbes. Published January 11, 2019. https://tinyurl.com/y763qtvr. Accessed December 11, 2020.

68 —— Carbon Disclosure Project (CDP) website. https://tinyurl.com/y3nks59x. Accessed December 11, 2020.

69 —— Ein kritischer Blick. Spektrum.de website. Published November 4, 2017. https://tinyurl.com/ydcb8l9y. Accessed December 11, 2020.

70 —— Roundtable on Sustainable Palm Oil (RSPO) website. https://rspo.org/about. Accessed December 11, 2020.

71 —— 8 things to know about palm oil. World Wide Fund for Nature (WWF) website. Published January 17, 2020. https://tinyurl.com/y2rubk9e. Accessed December 11, 2020.

72 —— WWF Statement on the 2020 Palm Oil Buyers Scorecard. World Wide Fund for Nature (WWF) website. Published January 17, 2020. https://tinyurl.com/y2r8evy5. Accessed December 11, 2020.

73 —— The Final Countdown. Now or never to reform the palm oil industry. Greenpeace report. 2018. https://tinyurl.com/yb7865hl. Accessed December 11, 2020.

74 —— Verbrannter Wald. Robin Wood Magazin. Nr. 139/4.2018. https://tinyurl.com/y4nywskc. Accessed December 11, 2020.

75 —— Brasiliens Wälder brennen wieder. So kannst Du helfen. Ecosia website. Published August 28, 2020. https://tinyurl.com/y49asypb. Accessed December 11, 2020.

76 —— Tobacco. World Health Organization (WHO) website. https://tinyurl.com/y66amzyz. Accessed December 11, 2020.

77 —— Obesity and overweight. World Health Organization (WHO) website, https://tinyurl.com/yxna6hdk. Accessed December 11, 2020.

78 —— Textile Exchange reports global production of organic cotton increases 56 percent to reach highest levels seen in eight years and growth is expected to continue. Press release. November 14, 2019. https://tinyurl.com/y8bt87j6. Accessed December 15, 2020.

79 —— Certified organic cotton. C&A website. https://tinyurl.com/y4pqpohv. Accessed October 26, 2020.

80 —— Hardcastle J. Why Most Companies Don't Link ESG Performance to Executive Pay. Environment & Energy Leader. Published January 29, 2016. https://tinyurl.com/y82vjcd3. Accessed December 15, 2020.

81 —— Human T. Most companies avoid putting ESG metrics in incentive plans. IR magazine. Published October 10, 2019. https://tinyurl.com/y9adyhjj. Accessed December 15, 2020.

82 —— Berg A. et al. Getting the most out of your sustainability program. McKinsey & Company website. Published August 1, 2015. https://tinyurl.com/ycyl8obb. Accessed December 15, 2020.

83 —— The Climate Pledge website. https://www.theclimatepledge.com/. Accessed December 15, 2020.

84 —— UK bans Shell ad of refinery sprouting flowers. NBC News. Published November 7, 2007. https://tinyurl.com/yd8ero4u. Accessed December 15, 2020.

85 —— Hough A. Watchdog says Shell recycling adverts "misleading". Reuters Business News. Published November 7, 2007. https://tinyurl.com/y9jsghs3. Accessed December 15, 2020.

86 —— Capri Sun website. https://tinyurl.com/ya8jeuta. Accessed December 15, 2020.

87 —— Positions. Novartis website. https://tinyurl.com/yczw58zr. Accessed December 15, 2020.

88 —— Sustainable supply. C&A website. https://tinyurl.com/y3txtn3y. Accessed December 15, 2020.

89 —— Propper S. Here are the companies that use social media best for sustainability marketing. GreenBiz. Published February 1, 2019. https://tinyurl.com/ybbfgaq8. Accessed December 15, 2020.

90 —— Nace T. We're Now At A Million Plastic Bottles Per Minute–91% Of Which Are Not Recycled. Forbes. Published July 26, 2017. https://tinyurl.com/y794vsnz. Accessed December 15, 2020.

91 —— Parker L. The world's plastic pollution crisis explained. National Geographic. Published June 7, 2019. https://tinyurl.com/y32e33jd. Accessed December 15, 2020.

92 —— Shampoo bottles made from recycled plastic. Head & Shoulders website. https://tinyurl.com/y5ns27e9. Accessed December 15, 2020.

93 —— Creating Smarter Packaging. Henkel website. https://tinyurl.com/ybhslrpf. Accessed December 15, 2020.

94 —— Adidas to launch new fabrics from recycled ocean plastic, polyester. Reuters at Yahoo News. Published January 28, 2020. https://tinyurl.com/yd9dt9rk. Accessed December 15, 2020.

95 —— Adidas website. https://tinyurl.com/yalcgboq. Accessed December 15, 2020.

96 —— Henkel integrates Social Plastic® in packaging for Beauty Care and Laundry & Home Care products. Press release. Published April 29, 2019. https://tinyurl.com/yd74pw7w. Accessed December 15, 2020.

97 —— Alcohol. World Health Organization (WHO). https://tinyurl.com/wv3mt9w. Accessed December 15, 2020.

98 —— Wieser L. These six young people are changing the future. Red Bull website. Published October 10, 2019. https://tinyurl.com/ydgbl43k. Accessed December 15, 2020.

99 —— Hockerts K., Wüstenhagen R. Greening Goliaths versus emerging Davids—Theorizing about the role of incumbents and new entrants in sustainable entrepreneurship. Journal of Business Venturing 25 (2010) 481-492. https://doi.org/10.1016/j.jbusvent.2009.07.005. Accessed November 15, 2020.

100 —— Katerva website. https://katerva.net/about/sustainable-innovation. Accessed December 15, 2020.

101 —— Business as Unusual. Yunus Social Business website. https://www.yunussb.com/business-as-unusual. Accessed December 15, 2020.

102 —— Förster U. Rügenwalder Mühle–auf ins Veggie-Universum. Food Service. Published November 3, 2018. https://tinyurl.com/y98km7mx. Accessed October 18, 2020.

103 —— Greenwashing: Die Rügenwalder Lüge. Vegan Info. Published September 16, 2019. https://veganinfo.blog/2019/09/16/greenwashing/. Accessed December 15, 2020.

104 —— Jobs and Development. World Bank website. https://tinyurl.com/y773jbzn. Accessed December 15, 2020.

105 —— The Sustainable Development Goals Report 2020. United Nations. 2020. https://tinyurl.com/y5rokp7c. Accessed October 20, 2020.

106 —— Decent Work and the 2030 Agenda for Sustainable Development. International Labour Organization (ILO) website. https://tinyurl.com/y3oheefy. Accessed October 20, 2020.

107 —— Schwartz J. et al. What is the future of work? Deloitte Insights. Published April 1, 2019. https://tinyurl.com/yccf9wua. Accessed December 15, 2020.

108 —— Few countries combine an environmentally sustainable footprint with decent work. International Labour Organization (ILO) website. https://tinyurl.com/y99u92vu. Accessed December 15, 2020.

109 —— UNDP Human Development Report 2015. https://tinyurl.com/y8f6qzkr. Accessed December 15, 2020.

110 —— The Future of Jobs. World Economic Forum (WEF) report. 2016. https://tinyurl.com/ht73pp3. Accessed December 15, 2020.

111 —— The Future of Jobs Report 2018. World Economic Forum (WEF). Published September 17, 2018. https://tinyurl.com/ybrup26m. Accessed December 15, 2020.

112 —— Nestlé needs YOUth. Nestlé website. https://tinyurl.com/ydhz3nnn. Accessed December 15, 2020.

113 —— Frequently Asked Questions. Access to Medicine Foundation website. https://tinyurl.com/y92p8zqx. Accessed December 15, 2020.

114 —— Access to Medicine Ranking. Access to Medicine Foundation website. https://tinyurl.com/y6tpv52m. Accessed December 15, 2020.

115 —— Lebensbaum website. https://tinyurl.com/y94foao2. Accessed October 28, 2020.

116 —— The Body Shop website. https://tinyurl.com/yd8jupbe. Accessed December 15, 2020.

117 —— What is a green job? International Labour Organization (ILO) website. Published April 13, 2016. https://tinyurl.com/ycp2k49g. Accessed December 15, 2020.

118 —— Smith C. Drucker on the 'bounded goodness' of corporate social responsibility. INSEAD KNOWLEDGE. Published January 25, 2010. https://tinyurl.com/ya2t8mrm. Accessed December 15, 2020.

119 —— The Ultimate List Of Charitable Giving Statistics For 2018. Nonprofit Source website. https://tinyurl.com/saeoynz. Accessed December 15, 2020.

120 ——Honda website. https://csr.honda.com/community/. Accessed November 20, 2020.

121 —— CEWE Sustainability Report 2019. 2020. https://tinyurl.com/yae9vvoy. Accessed October 28, 2020.

122 —— Charitable Giving. The Walt Disney Company website. https://tinyurl.com/ybh779w5. Accessed December 15, 2020.

123 —— Dow Promise. Dow Chemical website. Published May 29, 2018. https://tinyurl.com/y7rcr4vc. Accessed December 15, 2020.

124 —— BHP Community requirements. https://tinyurl.com/yb78kvqh. Accessed December 15, 2020. Published May 29, 2018.

125 —— BHP Our Community Contribution 2018/2019. https://tinyurl.com/y7wcrhcd. Accessed December 15, 2020.

126 —— Colleagues & Communities. Nordzucker website. https://tinyurl.com/y7avsoso. Accessed December 15, 2020.

127 —— Corporate Foundations: Getting started. Foundation Source. https://tinyurl.com/ybr4w9en. Accessed December 15, 2020.

128 —— The impact of multinationals in developing countries. World Benchmarking Alliance (WBA) report. Published May 7, 2020. https://tinyurl.com/yxmc7vb5. Accessed December 15, 2020.

129 —— Tata Strive website. https://tinyurl.com/yc7lblgx. Accessed December 15, 2020.

130 —— Biggest step in IKEA history taken to support integration of refugees by 2022. IKEA Foundation website. Published December 16, 2019. https://tinyurl.com/ybwcekla. Accessed December 15, 2020.

131 —— Corporations give for tsunami aid. NBC News. Published December 31, 2004. https://tinyurl.com/yc6xv3hj. Accessed December 15, 2020.

132 —— L'Oréal Takes Part: Our Response To Covid-19. L'Oréal website. https://tinyurl.com/yabj3gzb. Accessed December 15, 2020.

133 —— L'Oréal launches a Europe-wide coronavirus solidarity programme. Press release. Published March 18, 2020. https://tinyurl.com/y9m5yc7w. Accessed December 15, 2020.

134 —— Mars Commits $26M to Communities in COVID-19 Response. Press release. Published April 3, 2020. https://tinyurl.com/ybs3fhgu. Accessed December 15, 2020.

135 —— The Goldman Sachs COVID-19 Relief Fund & Employee Match Program. Goldman Sachs website. https://tinyurl.com/ybbahdce. Accessed December 15, 2020.

136 —— JBS Annual and Sustainability Report 2019. https://tinyurl.com/yd8tr7co. Accessed October 25, 2020.

137 —— JBS website. https://jbs.com.br/fazerobemfazbem/en. Accessed October 25, 2020.

138 —— Deutsche Post DHL Group provides disaster response team volunteers to support COVID-19 humanitarian relief efforts throughout the Americas. Press release. August 17, 2020. https://tinyurl.com/y82etoya. Accessed December 15, 2020.

139 —— Corona-Krise: Reemtsma finanziert Einzelunterkünfte für bis zu 250 Hamburger Obdachlose. Press release. April 8, 2020. https://tinyurl.com/ybxzsa8d. Accessed December 15, 2020.

140 —— Free online courses to help 25 million get new digital skills for the COVID-19 economy. Microsoft website. https://news.microsoft.com/skills/. Accessed December 15, 2020.

141 —— Corporate Responsibility. Emirates website. https://tinyurl.com/ybxeuce7. Accessed December 15, 2020.

142 —— Sponsorships. KLM website. https://tinyurl.com/ybl6nsp6. Accessed December 15, 2020.

143 —— ExxonMobil 2018 Sustainability Report Highlights. https://tinyurl.com/y7tlv53o. Accessed December 15, 2020.

144 —— Matching Gifts Program. Bank of America website. https://tinyurl.com/y7n9qamn. Accessed December 15, 2020.

145 —— Partnerships. World Wide Fund for Nature (WWF) website. https://www.worldwildlife.org/pages/partnerships. Accessed December 15, 2020.

146 —— Olivela website. https://www.olivela.com/charity. Accessed December 15, 2020.

147 —— The State of World Fisheries and Aquaculture 2020. Food and Agriculture Organization (FAO). https://doi.org/10.4060/ca9229en. Accessed November 28, 2020.

148 —— Bechtel website. https://www.bechtel.com/sustainability/. Accessed December 15, 2020.

149 —— Global and regional tourism performance. United Nations World Tourism Organization (UNWTO) website. https://tinyurl.com/qkka9ur. Accessed November 15, 2020.

150 —— The Travel & Tourism Competitiveness Report 2019. World Economic Forum (WEF). http://tinyurl.com/y8f9vb86. Accessed December 15, 2020.

151 —— International Tourism and Covid-19. United Nations World Tourism Organization (UNWTO) website. https://tinyurl.com/yaeuamld. Accessed October 25, 2020.

152 —— Country profile Maldives. United Nations World Tourism Organization (UNWTO) website. https://www.unwto.org/country-profile-inbound-tourism. Accessed October 25, 2020. 2019 data.

153 —— Parker L. Microplastics have moved into virtually every crevice on Earth. National Geographic. Published August 7, 2020. https://tinyurl.com/y7vp2fcs. Accessed December 15, 2020.

154 —— Maldives to Improve Solid Waste Management with World Bank Support. World Bank. Press release. Published June 23, 2017. https://tinyurl.com/y4cst4kn. Accessed December 15, 2020.

155 —— How Sustainable Tourism is impacting the Travel Industry. Travel Technology & Solutions website. Published October 24, 2016. https://tinyurl.com/y9zzzefs. Accessed December 15, 2020.

156 —— Remy N. et al. Style that's sustainable: A new fast-fashion formula. McKinsey & Company website. Published October 20, 2016. https://tinyurl.com/y3h3wsf3. Accessed December 15, 2020.

157 —— Auswirkungen der Corona-Krise: News aus Kambodscha. Kampagne für Saubere Kleidung website. Published July 21, 2020. https://tinyurl.com/ybqqjavt. Accessed December 15, 2020.

158 —— Fair Wear Foundation Annual Report 2019. https://tinyurl.com/y8pvt5f7. Accessed December 15, 2020.

159 —— Lohn zum Leben. Kampagne für Saubere Kleidung website. https://saubere-kleidung.de/lohn-zum-leben/. Accessed December 15, 2020.

160 —— A fair wage: A human right. International Labour Organization (ILO) website. Published December 9, 2013. https://tinyurl.com/y9w8ea69. Accessed December 15, 2020.

161 —— Country Profile Romania 2018. Clean Clothes Campaign. https://tinyurl.com/ycwtgl37. Accessed December 15, 2020.

162 —— Reichart E., Drew D. By the Numbers: The Economic, Social and Environmental Impacts of "Fast Fashion". World Resources Institute (WRI) website. Published January 10, 2019. https://tinyurl.com/y2rtwsdm. Accessed December 15, 2020.

163 —— Statista website. https://tinyurl.com/hy2skfk. Accessed December 15, 2020.

164 —— Smartphone Market Share. International Data Corporation (IDC) website. Updated December 15, 2020. https://tinyurl.com/ycab5kty. Accessed December 15, 2020.

165 —— Worldwide Tablet Shipments Continue to Decline in Q4 2019, According to IDC. International Data Corporation (IDC) website. Published January 30, 2020. https://tinyurl.com/y6vjdh5t. Accessed December 15, 2020.

166 —— Osterath B. Do you really need a new smartphone–or do you just want one? DW website. Published May 3, 2016. https://tinyurl.com/ydx4zeat. Accessed December 15, 2020.

167 —— Voss T. Let's make fair, sustainable mining the new normal. Fairphone website. Published June 23, 2020. https://tinyurl.com/yakxpj95. Accessed December 15, 2020.

168 —— This is what we die for: Human rights abuses in the Democratic Republic of the Congo power the global trade in cobalt. Amnesty International (AI) report. 2016. https://tinyurl.com/y4mzvfff. Accessed December 15, 2020.

169 —— Responsible Minerals Initiative website. https://tinyurl.com/ybwf83tm. Accessed October 28, 2020.

170 —— Fairphone website. https://www.fairphone.com/en/recycle-your-phone/. Accessed December 20, 2015.

171 —— Bevers B. Ist die Kreislaufwirtschaft eine Illusion? Deutschlandfunk website. Published January 9, 2019. https://tinyurl.com/y7wz8dll. Accessed December 15, 2020.

172 —— We're here. Fairphone Impact Value Report No. 1. 2018. https://tinyurl.com/yb5k3w6b. Accessed December 15, 2020.

173 —— Statista website. https://tinyurl.com/ych87kvp. Accessed December 15, 2020.

174 —— Wölbert C. Einmal Afrika und zurück: Streit über Elektroschrott-Export. heise online. Published July 30, 2014. https://tinyurl.com/ybqwey9r. Accessed December 15, 2020.

175 —— Turning the tide on e-waste in Nigeria protects the environment and creates safer jobs. World Economic Forum (WEF) website. Published June 27, 2019. https://tinyurl.com/y7hdr5c7. Accessed December 15, 2020.

176 —— RoHS website. https://www.rohsguide.com/rohs-faq.htm. Accessed December 15, 2020.

177 —— Kluijver J. Guest Blog: Collecting 75,000 scrap phones in Ghana. Fairphone website. Published August 21, 2014. https://tinyurl.com/ycp7j5lv. Accessed December 15, 2020.

178 —— EPEA website. https://tinyurl.com/y555uxkn. Accessed December 15, 2020.

179 —— Moss K. Here's What Could Go Wrong with the Circular Economy—and How to Keep it on Track. World Resources Institute (WRI) website. Published August 28, 2019. https://tinyurl.com/s8qmeyj. Accessed December 15, 2020.

180 —— Facts & Figures: The cold hard facts about overfishing. World Wide Fund for Nature (WWF) website. https://tinyurl.com/yclw36jr. Accessed December 15, 2020.

181 —— Guy A. Dispatch: In a city of 10 million, a community of traditional fishers faces big changes. Oceana website. Published May 18, 2018. https://tinyurl.com/y833zsyr. Accessed December 15, 2020.

182 —— Main ethical issues in fisheries. Food and Agriculture Organization (FAO) website. https://tinyurl.com/yd538qr6. Accessed December 15, 2020.

183 —— General situation of world fish stocks. Food and Agriculture Organization (FAO) analysis. https://tinyurl.com/yaos6te. Accessed December 15, 2020.

184 —— The Ocean Is Running Out of Fish. Here's the Alarming Math. YouTube video by Marinebio.org. https://tinyurl.com/ybefpns4. Accessed December 15, 2020.

185 —— Amoroso R.O. et al. Bottom trawl fishing footprints on the world's continental shelves. PNAS October 23, 2018 115 (43). E10275-E10282. First published October 8, 2018. https://doi.org/10.1073/pnas.1802379115. Accessed December 15, 2020.

186 —— To Manage Global Trawling Impact, Think Local. NOAA Fisheries website. Published February 12, 2020. https://tinyurl.com/yaca22h3. Accessed December 15, 2020.

187 —— Sustainable Management of Bycatch in Latin America and Caribbean Trawl Fisheries (REBYC-II LAC). Food and Agriculture Organization (FAO) website. http://www.fao.org/in-action/rebyc-2/en/. Accessed December 15, 2020.

188 —— Global Aquaculture Alliance website. https://tinyurl.com/yd6w2ak5. Accessed December 15, 2020.

189 —— Feeds for Aquaculture. NOOA Fisheries website. https://tinyurl.com/yddsvws7. Accessed December 15, 2020.

190 —— Sustainable Aquaculture Feed & Ingredients. Cargill website. https://tinyurl.com/y74t6xmk. Accessed December 15, 2020.

191 —— 3 Promising Alternative Feeds for Aquaculture. Rubicon Resources website. Published January 29, 2018. https://tinyurl.com/y7mzxtvd. Accessed December 15, 2020.

192 —— ASC Certification Update. Aquaculture Stewardship Council (ASC) website. https://www.asc-aqua.org/news/certification-update/. Accessed October 28, 2020.

193 —— How will cultured meat and meat alternatives disrupt the agricultural and food industry? AT Kearney report. 2019. https://tinyurl.com/y2gl9k6e. Accessed October 28, 2020.

194 —— Ritchie H. Food production is responsible for one-quarter of the world's greenhouse gas emissions. Our World in Data website. Published November 6, 2019. https://ourworldindata.org/food-ghg-emissions. Accessed December 15, 2020.

195 —— Electric Car Statistics in the US and Abroad. Last modified December 7, 2020. https://tinyurl.com/yd7wrgn2. Accessed December 15, 2020.

196 —— Bekker H. 2019 (Full Year) International: Worldwide Car Sales. Car Sales Statistic. Policy Advice. Published January 16, 2020. https://tinyurl.com/y7jat3oy. Accessed December 15, 2020.

197 —— Dawson Hoff V. 7 Eco-Friendly Fashion Labels To Know Now. Elle. Published April 22, 2014. https://tinyurl.com/y9dkm9uh. Accessed December 15, 2020.

198 —— Grüne Produkte in Deutschland 2017. Marktbeobachtungen für die Umweltpolitik. Umweltbundesamt (UBA). 2017. https://tinyurl.com/y6wntcxt. Accessed December 15, 2020. 2014 data.

199 —— Statista website. https://tinyurl.com/yan9gbyz. Accessed December 15, 2020. 2019 data.

200 —— Statista website. https://tinyurl.com/y7hfbxu8. Accessed December 15, 2020. 2019 data.

201 —— Vegan-Trend: Zahlen und Fakten zum Veggie-Markt. ProVeg International website, https://tinyurl.com/tv563kw. Accessed December 15, 2020.

202 —— Deutschland, wie es isst - der BMEL-Ernährungsreport 2019. Bundesministerium für Ernährung und Landwirtschaft (BMEL) website. https://tinyurl.com/y8y5q2vf. Accessed December 15, 2020.

203 —— Brand M. Fleisch in Deutschland. Statista website. Published October 4, 2018. https://tinyurl.com/y7cjj2bz. Accessed December 15, 2020.

204 —— Statista website. https://tinyurl.com/y4hvjubp. Accessed December 15, 2020.

205 —— Fleisch kostet Leben. PETA website. Published June 2006. https://tinyurl.com/y97u4vu9. Accessed December 15, 2020.

206 —— Halving food waste. Bundesregierung website. Published February 20, 2019. https://tinyurl.com/y4qltdj8. Accessed December 15, 2020.

207 —— Germany produces record amount of packaging waste. DW website. Published November 18, 2019. https://tinyurl.com/y78p9kyz. Accessed December 15, 2020.

208 —— Whelan T., Kronthal R. Research: Actually, Consumers Do Buy Sustainable Products. Harvard Business Review website. Published June 19, 2019. https://tinyurl.com/y74w2gap. Accessed December 15, 2020.

209 —— United Arab Emirate. Global Footprint Network website. Published November 18, 2015. https://tinyurl.com/y8g83ory. Accessed December 15, 2020.

210 —— Global Footprint Network website. https://tinyurl.com/y2naap4e. Accessed December 15, 2020.

211 —— Xu M. 5 charts show how your household drives up global greenhouse gas emissions. PBS website. Published September 21, 2019. https://tinyurl.com/yycsvscl. Accessed December 15, 2020.

212 —— Plastikatlas 2019. Heinrich Böll Stiftung / BUND report. Published July 3, 2019. https://tinyurl.com/yd2jwjfk. Accessed December 15, 2020.

213 —— Inequality.org website. https://inequality.org/facts/global-inequality/. Accessed December 15, 2020.

214 —— Rogers T.N. Here's how the ultrawealthy got even richer during the pandemic while millions of Americans faced job loss, hunger, and homelessness. Business Insider website. Published September 25, 2020. https://tinyurl.com/ya7hpgcy. Accessed December 15, 2020.

215 —— Widlitz S. Amazon: The Consumer's Everything In A Post Covid World. Forbes. Published July 31, 2020. https://tinyurl.com/y5b7puyy. Accessed December 15, 2020.

216 —— Sustainable selections: How socially responsible companies are turning a profit. Nielsen website. Published December 12, 2015. https://tinyurl.com/ya6z5tf5. Accessed December 15, 2020.

217 —— Adam D. Green idealists fail to make grade, says study. The Guardian. Published September 24, 2008. https://tinyurl.com/hzyweru. Accessed December 15, 2020.

218 —— Paris aligned annual carbon budget. Atmosfair website. https://tinyurl.com/y3jjxo6h. Accessed October 28, 2020.

219 —— Crippa, M. et al. Fossil CO2 emissions of all world countries. EDGAR Booklet. Published August 2019. Ispra: JRC SCIENCE FOR POLICY REPORT. 2019. https://tinyurl.com/y8usz45r. Accessed July 15, 2020. For an update of more recent data, see the report in reference 37.

220 —— Netflix und Co. verbrauchen weltweit gewaltige Strommengen. Tagesspiegel. Published March 3, 2019. https://tinyurl.com/y8enfdwf. Accessed December 15, 2020.

221 —— Spandick N. Wie viel CO2 produzieren wir durch Streaming und Googeln? Jetzt. Published August 21, 2019. https://tinyurl.com/y9vca5h4. Accessed December 15, 2020.

222 —— Kamiya G. Factcheck: What is the carbon footprint of streaming video on Netflix? CarbonBrief. Published February 25, 2020. https://tinyurl.com/w3wl9jq. Accessed December 15, 2020.

223 —— The carbon foodprint of 5 diets compared. Shrinkthatfootprint website. https://tinyurl.com/lqwjh8l. Accessed December 15, 2020.

224 —— Carbon and water footprint of diet choices. Animal Charity Evaluator website. Published September 2018. https://tinyurl.com/yaawquyb. Accessed December 15, 2020.

225 —— Fischer G. et al. Quantifying the land footprint of Germany and the EU using a hybrid accounting model. Umweltbundesamt (UBA) TEXTE 78/2017.

226 —— What If We Kept Our Cars Parked for Trips Less Than One Mile? Environmental Protection Agency (EPA) website. Published June 2015. https://tinyurl.com/yyaqfjxh. Accessed December 15, 2020.

227 —— Heimann S. Ökologisch reisen und nachhaltig mobil sein. co2online website. https://tinyurl.com/yafuedmg. Accessed December 15, 2020.

228 —— Petersen L. Small change: cutting your shower time by one minute. Ecopedia website. Published March 28, 2013. http://tinyurl.com/ybg53fdp. Accessed December 15, 2020.

229 —— Sustainability. Boston University website. http://tinyurl.com/n5rhlu7. Accessed December 15, 2020.

230 —— The diet that helps fight climate change. YouTube Video on Vox Channel. Published December 12, 2017. https://www.youtube.com/watch?v=nUnJQWO4YJY. Accessed October 28, 2020.

231 —— Ospina C. Cooling Your Home but Warming the Planet: How We Can Stop Air Conditioning from Worsening Climate Change. Climate Institute website. Published August 7, 2018. https://tinyurl.com/ydfbek3c. Accessed December 15, 2020.

232 —— Ambrose J. Air conditioning curbs could save years' worth of emissions – study. The Guardian. Published July 17, 2020. https://tinyurl.com/y3cghhrk. Accessed December 15, 2020.

233 —— Parker T. From Austria to Wales: The five best recycling countries of the world. NS Packaging. Published November 13, 2020. https://tinyurl.com/y8tbf6gb. Accessed December 15, 2020.

234 —— Top 10 Recycling Countries From Around The World. General Kinematik website. Published September 19, 2016. https://tinyurl.com/yconqcom. Accessed December 15, 2020.

235 —— Glas und Altglas. Umweltbundesamt (UBA) website. October 6, 2020. https://tinyurl.com/yd5o3czd. Accessed December 15, 2020.

236 —— Most Recyclable Materials & Recycling Rates Of Different Materials (Paper, Metals, Plastic etc.). Better Meets Reality. Published February 21, 2019 and updated August 31, 2020. https://tinyurl.com/yckb3zl3. Accessed December 15, 2020.

237 —— Taylor L. List of Materials That Are Recyclable. Sciencing website. Updated November 22, 2019. https://tinyurl.com/yddgddaa. Accessed December 15, 2020.

238 —— Steel is the World's Most Recycled Material. Steel Recycling Institute website. https://www.steelsustainability.org/recycling. Accessed December 15, 2020.

239 ——How Does U.S. Recycling Compare to the Rest of the World? Perfect Rubber Mulch website. https://tinyurl.com/y7re8let. Accessed December 15, 2020.

240 —— ABAL website. http://abal.org.br/en/statistical-information/. Accessed December 15, 2020.

241 —— Sorting waste paper at home for recycling. Recycling.com website. https://www.recycling.com/paper-recycling/. Accessed December 15, 2020.

242 —— Chinese Textile Recycling: The Night Is Darkest Just Before Sunrise. Global Recycling website. https://global-recycling.info/archives/3228. Accessed December 15, 2020.

243 —— International Recycling Group website. https://www.internationalrecycling.com/. Accessed December 15, 2020.

244 —— What is Atmosfair. Atmosfair website. https://tinyurl.com/y9769ger. Accessed December 15, 2020.

IMPRINT

Do you speak sustainability?
A personal navigator for corporate action

Copyright ©2020 by Anke Steinbach
All rights reserved.

Hardcover ISBN   978-3-9822766-8-7
Ebook ISBN   978-3-9822766-0-1

Publisher: Anke Steinbach

Cover design, interior design and implementation by Anna Beddig, www.annabeddig.de

Professional developmental editing by Marc Winkelmann, www.marcwinkelmann.de

English editing and proofreading by Nicole Cousins, ncousins@klusmannco.de

Photo back cover by Rachel Martin, @rachelmartinphoto

Printed in Germany: WirmachenDruck, www.wir-machen-druck.de

Printed on recycling paper

Climate-neutral printing: The CO2 emissions associated with the hardcover edition are offset by WIRmachenDRUCK. The offsetting project can be viewed with the ID on the website at www.climatepartner.de.

To order directly, order online at www.sustainavigator.com

Postal address: Anke Steinbach, CS Business Center, Am Kaiserkai 69, 20457 Hamburg, Germany

E-mail: hello@sustainavigator.com

With the sale of each book, 10% of the proceeds will be donated to initiatives promoting sustainability. The initiatives supported will be shared on social media.

@sustainavigator